Library of
Davidson College
VOID

THE BEST
OF THE
NAUTILUS

THE BEST OF THE NAUTILUS

A Bicentennial Anthology of American Conchology

EDITED BY
R. Tucker Abbott

greenville
delaware

THE BEST OF THE NAUTILUS

Copyright 1976 by R. Tucker Abbott. All rights reserved. No part of this publication may be reproduced, stored in a retrieval system, or transmitted, in any form or by any means, electronic, mechanical, photocopying, recording, or otherwise, without prior written permission of the editor or publishers, except in a magazine, scientific journal or newspaper article for purposes of review.

Library of Congress Catalog Card Number: 75-41628
International Standard Book Number: 0-915826-02-X

Warning: The advertisements reproduced in this book are merely of historical interest and do not represent living persons or products still available. Their mailing addresses are no longer valid. Information on the availability of back numbers and subscriptions to *The Nautilus* may be obtained by writing to: *The Nautilus*, Box 4208, Greenville, Delaware 19807.

Published by *American Malacologists*,
P. O. Box 4208, Greenville, Delaware 19807
PRINTED AND BOUND IN THE UNITED STATES OF AMERICA.

DEDICATED TO

Bill Clench

Humble giant of malacology who has produced many of today's leaders in the field, and has long been a beloved inspiration to a nation of dedicated shell collectors.

Frontispiece

Some Shells Featured in *The Nautilus*

1. *Canthyria spinosa* Lea. Spiny Unio from the Altamaha River, Georgia.
2. A young *Cassis madagascariensis* Lamarck from off North Carolina.
3. The Junonia, *Scaphella junonia* Lamarck, from Sanibel Id., Florida.
4. *Tiphobia horei* E. A. Smith, a deep water snail from Lake Tanganyika, Africa.
5. *Io fluvialis (Say)*, a Tennessee River snail, now rare.
6. and 11. Cuban bush snails, *Polymita picta* (Born), from Oriente.
7. Interior view of the Abalone, *Haliotis kamtschatkana* Jonas.
8. Northern Pond Snail, *Lymnaea stagnalis* (Linnaeus), U. S.
9. White-lipped Woodland Snail, *Triodopsis albolabris* (Say), U.S.
10. Lemon Yellow Pecten, *Aequipecten muscosus* (Wood), Florida.

THE BEST OF THE NAUTILUS

CONTENTS

Introductions	1
Early Collecting on the Atlantic Coast	5
Pioneering on the Pacific Coast	41
Tracking the Terrestrials	83
Fording for Fresh-water Shells	135
Passions of the Pearly Mussel	159
Collecting Abroad	177
Departed Friends	231
Of Shoes – and Ships – and Sealing-wax	263
Salute to The Nautilus	275

Vol. I.   **No. 1.**

This is the first issue of "THE CONCHOLOGIST'S EXCHANGE." As encouragement is received it will assume the form of a printed sheet with columns for "Exchanges in Mollusca," "New Localities," "Answers to Correspondents," &c. This, our first number, has been sent to 500 Conchologists. Subscription price, 25 cents per annum, post paid. Exchanges of 20 words, 10 cents; for each additional 10 words the charge will be 5 cents. The Conchologist's Exchange will be issued semi-monthly, and will endeavour to become a cheap and useful medium for the exchange of those most beautiful productions of nature—"The Mollusks."

EXCHANGES FOR MOLLUSCA ONLY.

CYPRÆA erosa, L. lynx, L.
CERITHIUM, maculosum, Kien. eburneum Brug.
CYCLOSTOMA sulcatum, Lam. elegans Mull.
LYMNÆA zebra Tryon.
STROMBINA bicanalifera Sby. Fissurella volcano. Rve. Columbella fulgurans Lam.
    Prof. D. S. SHELDON,
        Davenport, Ia

SUCCINEA putris L.
HELIX arbustorum L.
    " nemoralis L
    " ericetorum, Mull.
    " rotundato, Mull.
    " lapicida L. cellaria Mull
Pupa muscorum, L.
Cionella subcylindriea.
        E LEHNERT,
            Washington, D. C.

GONIOBASIS simplex, Say. carinifera, Lam. bella, Con. perangulata, Con. sordida Lea symmetrica, Hald ebenum, Lea
Melantho subsolida Anth.
Unio-rubiginosus Lea. pustulosus Lea., gracilis, Barnes
        W. A. MARSH,
            Aledo, Illinois.

HELIX albolabris, Say. alternata, Say. clausa, Say. elevata, Say. fallax, Say. hirsuta, Say. inflecta, Say. solitaria, Say. monodon. Rackett. Sayii. Binney. Pupa armifera, Say. corticaria, Say. Fallax, Say. Unio elegans Lea; lachrymosus, Lea; parvus, Barnes
    EDWARD A. ENOS,
        Connersville, Indiana.

NASSA fossata Gld.
Purpura saxicola Val.
Amycla gausapata Gas.
Adula falcata Gld.
Acmæa spectrum Esch.
scabra, Nutt. pelta, Esch.
Hipponyx cranioides, Carp.
        G. W. PUTEBAUGH,
            Greenfield, Indiana.

American and Foreign Unionidae for exchange.
Send for list.

No responsibility will be assumed for the standing of the above parties.

Address, **WILLIAM D. AVERELL**, Proprietor,
CHESTNUT HILL, PHILADELPHIA.

The *Nautilus* had its beginnings as "The Conchologist's Exchange" in July 1886

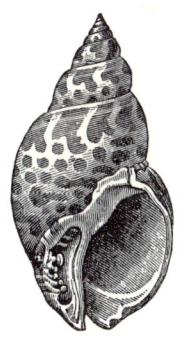

INTRODUCTION

    This is a contribution to the bicentennial celebration of the founding of our United States of America. It is a small and isolated contribution, but I think it may have considerable value, and certainly a great deal of interest, to Americans working in the field of malacology or studying conchology.
    This anthology is a carefully selected collection of the most interesting and significant articles that appeared in the first forty years of the publication of The Nautilus, America's oldest quarterly journal devoted to the interests of conchologists. Born in 1886 in Philadelphia, this journal has lived through 90 years of American science in times of boisterous pioneering, the confident calms of the early 1900's, the disturbing years of World War I and II, and the lean years of The Depression of the 1930's.

Few people have had the privilege and forbearance to have read the entire works published in the quarterly journal, The Nautilus. Today, there are probably no more than sixty complete sets of this periodical in the United States. Abroad there are less than half this number. A goodly part of the journal is devoted to dry, scientific descriptions of new species and to esoteric discussion of the physiology or biostatistics of land and fresh-water mollusks. However, a surprisingly large portion of the journal, particularly during the first forty years, is replete with charmingly written and very informative articles on collecting shells, as well as many valuable accounts of environmental conditions at the turn of this century.

Recurring throughout its pages over the years are two underlying themes, seldom expressed openly and directly, but always hauntingly present in the writings of such outstanding journalists as James Ferriss, and William S. Strode – firstly, the love of the quest for knowledge and the sheer joy of discovery, and, secondly, the almost pathetic realization of the impending doom of our natural environment. Time and again these early explorers and naturalists recorded major examples of the destruction of the fauna of many of our lakes, rivers, beaches and forests. In a sense, The Nautilus *is an ecological sampling of the industrial history of America. It is also a prelude of worst things to come unless society comes to grips with our ecological problems. This may be a time for Americans to celebrate, and certainly an occasion to appreciate, the sacrifices and labors of those who preceded us in the building of the United States and its freedoms, but perhaps it is more properly a time to consider changing our ways of unbridled waste and environmental suicide.*

On the more pleasant side there is in store for you many an hour of entertaining and nostalgic reading. Where possible I have tried to put various articles in their modern day perspective, to give a brief history of the writers and in some cases to bring the subject up to date.

Come with me, in the pages of The Nautilus, *down the upper Mississippi River in a boat on a unio-hunting trip*

Introduction

with Frank C. Baker in 1902; or venture with the Australian malacologist, Charles Hedley on his first helmet dive on a coral reef in 1894; or shell collecting in Alaska in 1897 during the Klondike gold rush with a 100-pound back-pack at 3,000 feet. Join James Ferriss and George Clapp at Clingman's Dome in the Great Smoky Mountains of Tennessee, and smell the wood smoke and breakfast bacon as they recount their days of land shelling during 1899. Read about land snails that stopped a train, and pleurocerid snails that clogged the water mains of Mark Twain's Hannibal, Missouri, in 1895. Laugh at Abe Loche, a policeman of Atlantic City who was felled by a clam dropped on his head by a seagull in 1925. Your admiration for America's most popular shell collector will increase tenfold when you read Henry A. Pilsbry's fascinating and touching obituary of James H. Ferriss (1849-1926).

The Nautilus, like so many scientific journals, is a nonprofit enterprise kept alive solely by the subscribers and the unpaid editors. Henry A. Pilsbry, Charles W. Johnson, H. Burrington Baker, and now myself, have collectively given 175 years of volunteer service to the cause of American malacology. Mrs. Bernadine ("Bunny") Baker has unstintingly contributed 38 years of her time to managing the subscriptions and business matters of the journal. I have faith that the members of the younger generation will step forward when it becomes their time to carry on this journal. The entire royalties from this anthology are being contributed to the improvement and continuation of The Nautilus.

R. Tucker Abbott, Editor
Greenville, Delaware, 1976

THE NAUTILUS.

Vol. III. MAY, 1889. No. 1.

INTRODUCTION.

THE publishers of THE NAUTILUS feel that no explanation of their object in offering this journal to the scientific public is necessary. The need of an American publication devoted especially to the interests of Conchologists is felt throughout the country. One of the greatest difficulties which the student of science has to overcome is found in the scattered and fragmentary character of scientific literature. The "Proceedings" or "Transactions" of a hundred societies, and the pages of innumerable journals must be searched through before one can be certain that a given fact or observation has or has not been recorded.

The simplest way to better this condition of things will be to limit by some means the number of publications in which a certain subject is likely to be treated upon; and this is most easily done by establishing journals devoted to special branches of science. It is the aim of THE NAUTILUS to afford such a medium for all who are interested in studying the Mollusca; and to this end the co-operation of all friends of science is solicited.

All subscribers to the *Conchologists' Exchange* (of which this paper is the successor) will be credited on the books of THE NAUTILUS with the amounts due them upon the suspension of that journal. All subscribers will be allowed one insertion of twenty-five words in the Exchange Column, free of charge.

Early Collecting on the Atlantic Coast

The joys and excitement of marine collecting are no better described than in Charles T. Simpson's 1897 article on "The Janthinas". This account was quoted in full in Julia Roger's famous The Shell Book in 1908 during her discussion of Violet Snails. The 1892 account of collecting in Tampa Bay is no less interesting. Going northward, further articles on collecting are contributed by C. W. Johnson at St. Augustine, John Ford along the New Jersey coast, A. Jacot in New York and several old-time collectors in New England. A brief biographical sketch of Simpson (1846-1932) is given in the chapter on "Tracking the Terrestrials". John Ford's life (1827-1910) is told in the chapter on "Departed Friends."

Many of the scientific names of the marine mollusks used during the early part of this century will be unfamiliar to many of today's conchologists. The changing of these names over the years is a natural result of additional research and new knowledge. The old names and their modern equivalents are to be found in the second edition of Abbott's American Seashells (Van Nostrand/Reinhold, N. Y., 1974).

THE IANTHINAS.

BY CHARLES T. SIMPSON.

The Ianthinas, or violet snails, live gregarious in the open seas of the tropics, and float by means of a raft composed of vesicles filled with air, which cannot be withdrawn into the shell. Sometimes they are carried by winds and currents into the seas of temperate regions, and their shells have been found along the shores of our own country as far north as New England. I had collected for many years and in many countries, but had never found, perhaps, more than a dozen dead, broken shells. In January, 1883, I was on a large schooner bound for Spanish Honduras, and we stopped at Key West, where I spent one of the most delightful weeks of my life gathering Cylindrellas, Chondropomas, Cerions, *Helicina orbiculata*, and the beautiful Orthalicus, Liguus, and *Bulimulus multilineatus* in the thick, thorny, tropical scrub, or Strombs and bright Tellinas and blending Neritas and a hundred other interesting forms along the south shore. We were to sail about noon on Sunday, but I could not resist the temptation to take one last look at the places where I had spent so many happy hours, so after breakfast I wandered through the city and out to the beach.

Before I reached it I noticed that as far as the eye could see, it was a mass of the most intense, glowing violet color, and on coming up to it was astonished to find that this color came from untold millions of Ianthina, which had been washed up in the night, for when I had left the beach the evening before at dusk, not one was to be seen. To say that they lined the shore gives no idea of the real truth. Everywhere, from below low water to highest tide mark they were piled up, in most places, over shoe-top deep, and in the hollows of rocks one could have waded in among them up to his knees—shell, animal and float all of a vivid purple, the richness of which soon fades, to a great extent, in dead shells or preserved specimens. They were all dead—a kind of slimy mass—and they somehow looked pitiful.

There had been no storm, nothing but an ordinary breeze blowing up from the south, and it is probable that an immense school had been drifted along, and where they struck the island, some five miles in length, every one within that distance was stranded. I had brought no basket or sack or anything to collect in, but I could not bear to go away and leave that vast bed of treasures without taking at least a few with me. I searched in vain for a box or tin can or a piece of canvas, but could find absolutely nothing, not even a scrap of paper. I took out my handkerchief, knotted the corners, and tried to pull out the animals from the shells, but the whole mass was so slippery and the shells were so frail that the latter invariably broke, so I filled it with shells, animals and all, as many as it would hold. Then I took off my straw hat and filled it, and that would not satisfy me, for as I wandered along I found so many fine specimens that I began to put them into my pockets, and I did not leave the shore until every pocket was bursting full. I had on a linen coat and white duck pants; the day was hot and it seemed to me that those Ianthinas melted. In a little while streaks of glowing violet began to show down my clothes; I felt a clammy, wet, uncomfortable, feeling clear through to my skin, and my shoes were filled with purple liquid. By the time I reached the city I looked like an Indian in war paint, and I have no doubt that the people of Key West, who were just going to church, thought I was a lunatic, and perhaps they were not far from right. At last I reached the schooner, took off and threw away my suit, which was utterly ruined, and got my precious mollusks into sea water to soak, although at least half of them were broken, yet, when I cleaned them out, I had the satisfaction of counting up over 2,000 good shells.

COLLECTING NOTES.

BY CHAS. T. SIMPSON.

During a brief vacation last Christmas, Mr. John B. Henderson, Jr., of Washington, and the writer made a flying visit to the west coast of Florida, in the vicinity of Tampa Bay, for the purpose of collecting shells; and I have thought that perhaps a few notes on our work might be of interest to the readers of the NAUTILUS. The country throughout this region consists of ordinary sandy pine land, interspersed here and there with ponds and hammock or hardwood tracts of from an acre or so to several miles in extent. This region in general is one of the flattest on the globe, and as a consequence the sea is in most places quite shallow and thousands of acres of mud flats are often laid bare at very low tides or during "Northers," affording wonderfully rich collecting grounds for the naturalist. We fitted ourselves out with a five or six ton sail-boat accompanied by a skipper and a good-natured cook, and with two weeks provisions, a gallon of alcohol, a dredge, and several large note books which were to be filled with original observations and discoveries, we sailed away as eager for adventures as Lord Bateman.

I want to say to anyone who attempts to collect marine shells or animals, that first and foremost it is all important to use the dredge. This implement is so simple, so easily constructed, and is so efficient that the merest tyro never ought to try to get along without it. A full description of one and its mode of working can be found in Woodward's Manual of Conchology, and one that brings the matter down to date will soon appear in a forthcoming paper by Dr. Dall on collecting. We threw overboard our dredge in the warm bright waters of Tampa Bay as the boat was brought up into the wind, and awaited results. There is a certain kind of excitement about the operation; the jar and tremble of the rope as the implement—far down out of sight—scrapes over the bottom, gathering in the treasures of the deep, produces a sensation akin to that which an angler feels when he gets a bite, or a sportsman when he sights game and "draws a bead." And this feeling reaches a fever heat when the dredge is hoisted slowly, leaving a cloudy wake in the water, and its contents are dumped into the screen.

Starfishes, echini, perhaps a big horseshoe crab or two, and, mingled with living mollusks and fishes there may probably be dead shells inhabited by various forms of hermit crabs, fish, sea-worms and a dozen other kinds of life, many of which may be puzzling even to an experienced naturalist. There is something wonderful about all this, and entirely different from shore collecting; the animals are taken in their homes, caught in the very act of carrying on their ordinary avocations, and it is not to be wondered at that they seem to have a kind of surprised appearance when they are tumbled out indiscriminately on deck. There is always an element of uncertainty about dredging that furnishes a mild excitement akin to that of gambling. One throw, or a half dozen in succession, may turn out to be "water hauls," bringing up nothing but mud or possibly sea urchins, and the "just once more before we go away" may bring up half a hundred species, some of them rare, and all desirable.

The vicinity of Tampa Bay is rich in marine species and is classic ground to the conchologist and the collector, it having been worked over by Agassiz, Conrad, Stimpson, Spinner, and other noted men who have passed on, and Drs. Stearns and Dall, Velie, Calkins, and others who are still with us; and often a run along the shores of some of the outer keys, or about the muddy, sandy bays, will reveal shells enough to turn the head of even a steady-going experienced conchologist. And at such times it always happens that when the collector gets every bucket, and sack, and basket, and both hands loaded down to the last limit with things that are good enough in all conscience, and is miles away from his boat, he begins to run upon numbers of such marvelously rare and beautiful things that he is tempted to throw away every thing he already has and begin entirely anew.

Prof. Hornaday spoke the truth when he said that "the collector's life is a constant race for specimens." In the few brief days we had at our command we felt that we must "make hay while the sun shone." But dredging, though very delightful at first, when followed up for eight or ten hours consecutively gets to be a good deal like work, and hard, heavy, wet work too. So we did what I should advise all collectors in similar circumstances to do; we went ashore during low tides and searched sometimes the sandy bays, the limited areas of rocky shore to be found about that region, or the open beaches; and during high tide we dredged. One rocky bed laid bare at low tide in Terraciea Bay was marvelously rich in *Tritonidea tincta, Cerithium floridanum, Semele reticulata, Murex nuceus, Cumingia tellinoidea, Nassa consensa, Muricidea multangula, Urosalpinx perrugatus,* and some other forms not often found on the sand.

Mrs. Mean's injunction "while yer a gittin' git a plenty" especially applies to the collector. One is prone when he sees anything in great abundance to feel as though it was very common and was hardly worth taking. Even the sight of a very rare mollusk in quantities somehow lowers its value in one's estimation. But the old collector who has let such chances go a few times, and afterwards where his entire stock of the same thing has run out, regrets his folly, learns to take all he can get of anything that is good. One may find a species thrown up to-day by millions on a certain shore, in excellent condition, and the next week, and for years afterwards, he may not run across a dozen individuals of the same. I had lived near Tampa Bay for four years and collected industriously, but throughout my whole residence I never found a hundred specimens of *Olivella mutica*, dead or alive. During our visit the dredge brought them up living, glittering like dew drops, by the handfuls. We dredged over and over the ground on which I once obtained in quantities of *Venericardia tridentata* and *flabella*, *Parastarte triquetra*, and *Pandora trilineata*, and scarcely found a specimen, while on the same ground we got a great many Tubonillas, an abundance of *Conus peali*, and a half bushel or more *Arca transversa*, not a specimen of which I had ever found there before; and on a sand flat that used to gladden my eyes with *Conus floridanus* not a single one could we find.

The shell mounds—the *Kitchen middens* of prehistoric tribes—are usually overgrown with tropical scrub, and are rich in land shells as well as mosquitoes and sand flies; and on one of these at Shaw's Point I rediscovered *Zonites dallianus* which I first found there three years ago, and, at the time, supposed to be the very different *Zonites minusculus*. In places the brackish water was swarming with *Cerithium minimum*, and *muscarium*, *Melampus coffea*, *Macoma constricta*, *Natica duplicata* and its companion *Melongena corona*, *Lucina Jamaicensis*, *Cerithidea scalariformis*, *Modiola plicatula*, var. *semicostata*, *Mytilus hamatus*, and the two Cyrenas, *floridana* and *carolinensis*. The ponds were alive with *Physa heterostropta* var. *pomilia*, *Succinea luteola*, which seems about as completely aquatic as any of the pond snails, *Planorbis tumidus*, which is a form of the protean and widely distributed *trivolvis;* and on the keys several of the Polygyras were abundant.

Our ten days of collecting came to an end all too soon, for although we had worked very hard and gathered in some 200 species and perhaps 25,000 specimens, we had not had time to write a half dozen notes, and we had only made a beginning at what we wanted to accomplish. We packed our material and bid good-by reluctantly to the land of palmettoes, warm breezes and sparkling waters.

Vol. XXXIII   JULY, 1919   No. 1.

AN OLD COLLECTING GROUND REVISITED.

BY CHARLES W. JOHNSON.

Charles Willison Johnson was one of the outstanding naturalists of the United States at the turn of this century. Although well-known as a malacologist and paleontologist, he was also recognized as a leading authority on the Diptera, or two-winged flies. Johnson was very popular with his colleagues and especially with young students, such as William J. Clench and the famed Harvard anatomist, Frederic T. Lewis. The latter wrote of Johnson:
"The door of Mr. Johnson's room was always open. Whenever any one, - school-boy, expert, old friend or stranger - crossed the threshold, his work was laid aside. His time and wealth of information were at every one's disposal. Often I have visited there, and never without learning much and developing a taste for more, as with the utmost simplicity Mr. Johnson drew upon his boundless lore of insects and mollusks. For him there was no idle time. When absent from the Museum he was always off collecting."

Born in Morris Plains, New Jersey, on October 26, 1863, Johnson completed his education in public and private schools at Morristown. In 1880, at the age of 17 he moved to St. Augustine, Florida, where he began making extensive collections of fossil shells throughout the state. There he soon met Dr. William H. Dall of the National Museum and Mr. Joseph Willcox, a Trustee of the Wagner Free Institute of Philadelphia. Both of these scholars were also interested in the Tertiary fauna of Florida, and in the latter part of 1888 we find Johnson, at 25, installed as Curator of the Wagner Free Institute.

Johnson was very active among the circle of great scientists in Philadelphia. He was soon made an honorary curator of fossil mollusks at the Academy of Natural Sciences of Philadelphia, and in 1890 joined H. A. Pilsbry's Nautilus *venture as business manager. In 1897, he married Carrie W. Ford, daughter of the conchologist, John Ford (1827-1910). She was one of the young ladies who painted shell plates for George W. Tryon, Jr. (1838-1888). In 1903 he moved to Boston where he became the principal Curator of the Boston Society of Natural History. In 1910, he organized the Boston Malacological Club. Of the 230 articles and reviews contributed to* The Nautilus *by Johnson, his better known ones dealt with the Olividae shells and collecting in St. Augustine. Johnson died in Brookline, Mass., a year after his wife's death, on July 19, 1932.*

While a resident of St. Augustine, Florida, from 1880–88, I made a careful study of the mollusca of the harbor and vicinity. The habits of the various species and the factors governing their distribution, which in many cases was much restricted, especially appealed to me. With these facts in mind it was with great interest that I visited the old city after an absence of thirty-one years. Time and the ever-shifting sands have played sad havoc with many of my old collecting grounds, and I looked in vain for some of the rarer species.

The accompanying maps can give only a general idea of the changes that have taken place.[1] The "Lagoon" of the eighties is gone and there are now two inlets with about the same depth of water on each bar according to the government chart, survey of 1910, although I was told that the southern channel has now much less water on the bar than the other. Marsh Island at the mouth of Hospital Creek is also gone, and the sand bar that was formerly only east of the island now extends to the fort. There is no trace of the site of the old Spanish lighthouse,

[1] Figure 1 shows the harbor and vicinity about 1883, before the St. Sebastian marsh was filled, also the approximate positions of the "Lagoon" and Marsh Island.

Early Collecting on the Atlantic Coast   13

which was probably at the extreme end of the now exposed ledge of coquina and about 200 feet below the present high-

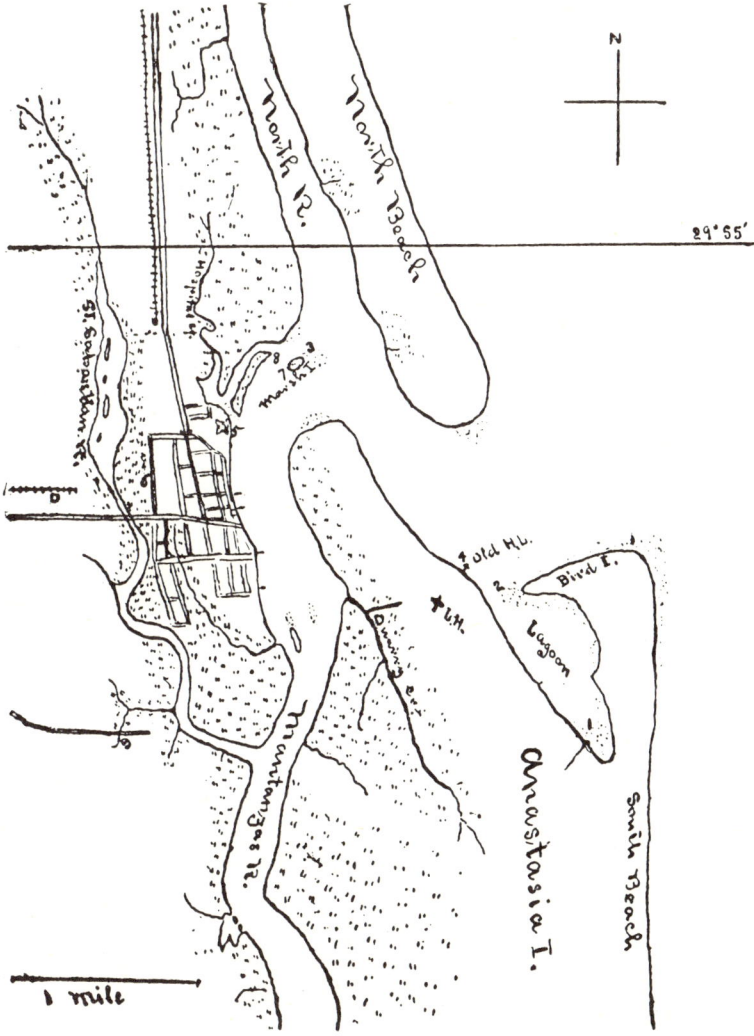

Fig. 1. St. Augustine, about 1883.

water mark. The government has endeavored to prevent the wearing away of this portion of Anastasia Island by construct-

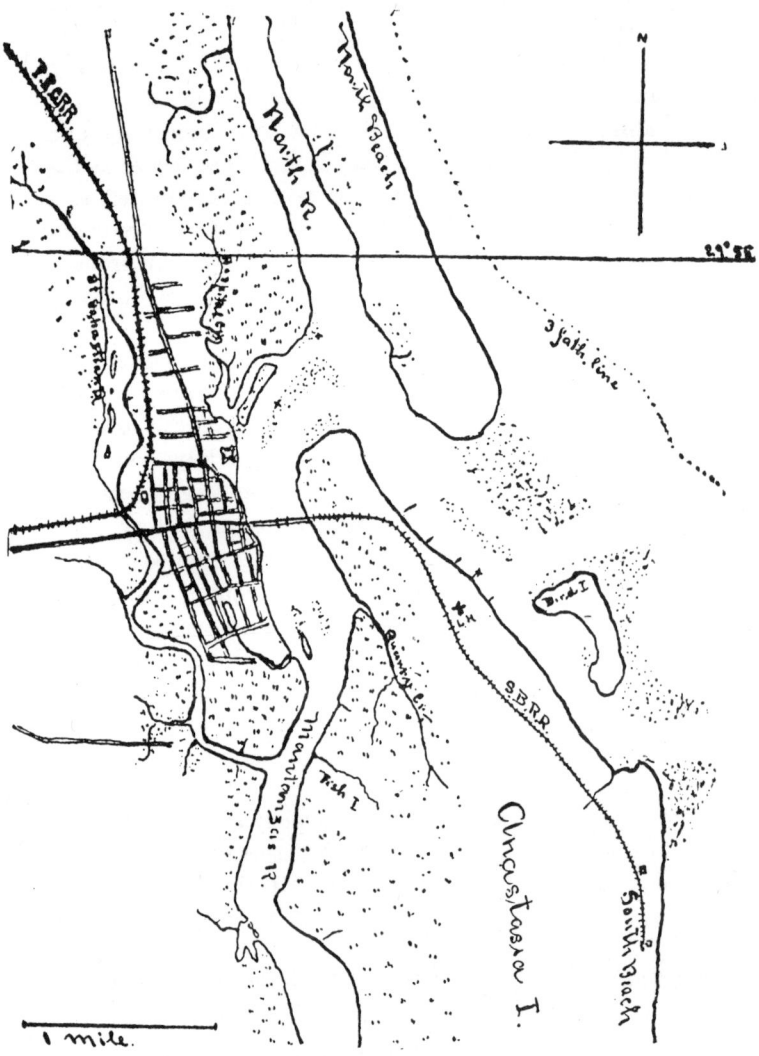

Fig. 2. St. Augustine at present.

ing four jetties, one below and three above the ledge referred to, but the erosion apparently still continues. The following notes are based chiefly on those species that were very limited in their distribution and which therefore may have become locally extirpated by the many changes affecting their environments. A list of about 200 species published by the writer in 1890 [1] forms the basis of some of the following remarks.

Macrocallista nimbosa Sol. (1) This is the *Callista gigantea* Gmel. of my list. It was found only in the shoal water at the head of the "Lagoon," seeming to prefer the quiet water, as I never found a trace of it on the ocean beach. At most only six or eight specimens were found, and many of these were broken, probably by the large ray or "clam cracker" as the butterfly ray (*Pteroplatea maclura*) is called by the fishermen.

Donax obesa d'Orb. (2) This little chunky species was formerly common on the sand bars at the mouth of the "Lagoon," where there was a slight shifting of the sand at every tide. The larger species, *Donax variabilis* Say, was (and probably is) exceedingly abundant on the ocean beaches, especially the "South beach." I was quite amused at Daytona to hear the popular name of "coquina" applied to this shell, and one young man talking about "coquina bouillon." While this is entirely proper, as the Spanish word coquina means, broadly speaking, shell-fish, the name has become so generally used for the shell-rock (often made up largely of this species) that at first it sounded like pretty hard diet. I am sorry that opportunity did not permit my getting a large series of this species including the young, as I should have liked to have made some comparisons of the young of *D. variabilis* with that of the typical or more northern *D. fossor* Say. As I remember I could never satisfactorily separate the two forms at St. Augustine and omitted the latter from my list, although it is recorded from the entire coast of Florida and westward to Texas. Mazyck in his "Catalog of Mollusca of South Carolina," says of *D. fossor*, "very rare, Sullivan Island."

[1] An Annotated List of the Shells of St. Augustine, Florida, THE NAUTILUS, vol. iii, pp. 103, 114 and 137, vol. iv, pp. 4–6.

On one visit to the South beach I found it literally strewn with perfect specimens of *Divaricella quadrisulcata* d'Orb. (*Lucina dentata* of my list), but never again did I find them in such numbers. At another time quantities of an Arca referred to in my list as *Arca americana* Gray, were found. It is more elongated than those found in the harbor, with a thinner and lighter brown periostracum, and probably represents only a variety or ocean form of *A. campechiensis* Dillw. (Arca pexata Say).

Lucina philippiana Reeve (*Loripes edentula* of my list). Large single valves were frequently found and occasionally at the mouth of Hospital creek shells were found in the mud with both valves intact, but like the *Phacoides filosa* Conr. in Portland harbor, Maine, always dead. It may also be of interest to note that two specimens of *Solemya velum* Say, and a few single valves of *Mya arenaria* were found on the north beach, the most southern records for the species.

Panopea bitruncata Conr. (3) This large and variable shell which was referred to in my list under both *Glycimeris reflexa* Say, and *G. americana* Conr., was later the subject of a paper by the writer [1] in which the synonymy was straightened out, and a fine specimen found on the bar east of Marsh island was figured. Common in the pliocene of the Caloosahatchie, but recent examples are exceedingly rare. Living deep in the mud they are difficult to obtain, unless on rare occasions extensive harbor dredgings might bring some to the surface. They are also very apt to be destroyed by changes such as encroaching sand bars, sedimentary deposits, and harbor pollution.

The rocks that represented the ruins of the old Spanish lighthouse (the tower of which fell in June, 1880, while the keeper's house had fallen several years before) were a favorite place for *Thais floridana* Conr. (*Purpura haemostoma* var. *floridana* of my list). During my recent visit I failed to find a living specimen of this species either on the ledge or jetties, but the tides were not very low and it may be that they could have been found at a lower tide. On all of the rocks including the

[1] THE NAUTILUS, vol. 18, pp. 73–75, pl. 4, 1904.

jetties were great numbers of *Siphonaria naufragum* Stearns (*S. lineolata* d'Orb.). One thing that seemed to impress me more than when I lived there, was the great abundance of oysters on all the rocks, even around the water battery of the fort and also on the piling. In speaking to an old friend regarding the matter, he said he thought that around the fort it was due to cleaning off the rocks a few years ago, thus presenting a clean surface for the young to cling to. This array of bristling oysters around the water battery of the fort deterred me from a hunt for *Nerita peloronta* and *N. versicolor* (5), three living specimens of which I found there together with *Litorina angulifera*, being the most northern record for the three species.

Cerithidea scalariformis Say (6). The only place that I ever found this species at St. Augustine was in the more sandy portion of the marsh west of the city between King street and Orange street, not far from where the Y. M. C. A. building now stands. The filling-in of the marsh has probably locally extirpated this species. Another related species *Cerithium floridanum* Mörch (7), *C. atratum* of my list, was also restricted to a small area, an old oyster bed at the west end of Marsh island. This is now a sand bar and the species may now be entirely absent in the harbor. At the latter place I also found my only living example of *Murex fulvescens* Sowb.

At the mouth of Hospital creek was a large patch of the grass-like Gorgonia—*Leptogorgia virgulata*. On this lived the little *Simnia uniplicata* Sowb. 8 (*Ovula uniplicata*), as the Gorgonia varied in color so did the shells of the *Simnia*, agreeing in color with the bunch of Gorgonia on which they were found— either white, light-yellow, orange or pink. On one occasion while hunting for *Simnia* a conspicuous object attracted my attention, its flesh-colored mantle with irregular blackish markings was very striking, and as it contracted I found I had a *Cyphoma gibbosa* Linn. (*Ovula gibbosa*), common to the West Indies. For some time I wondered why the animal of this shell should be so very conspicuous; then the thought occurred to me that in more southern waters probably most of them live on the "sea-fans" (*Rhipidogorgia flabellum*) and with their reticulated structure as a background the animals would be scarcely distinguishable, like the Sargassum fish (*Pterophryne histrio*) in the gulf-weed (Sargassum).

Cyrena carolinensis Bosc. (9). In making a bridge across a small branch of the St. Sebastian River I first discovered this species. It was a large and interesting variety in which the umbones were unusually high, many of the specimens closely resembling in size and form the figure of *C. olivacea* Carp. from Mexico, as given by Prime (Monograph American Corbiculidae, p. 17, fig. 12, 1865). Although the tide flowed freely up the little creek, there was at low tide a small stream of fresh water even at the driest time. At the junction of this little stream and the high ground there was a small colony of *Neritina lineolata* Lam. (*N. reclivata* Say). I looked in vain for this species during my recent visit, nor did I find *Cyrena* near the little bridge, but it may still exist in other parts of the stream which time did not permit me to examine thoroughly. At the mouth of Pellican creek near the Matanzas Inlet was a colony of *Neritina virginea* Linn. They were the olive-green or more brackish water type and probably represent the most northern limit of this species on the Atlantic coast. About seven miles south of Matanzas Inlet was a large cypress swamp in what was known as "Bike's Hammock," here was found *Ampullaria depressa* var. *hopetonensis* Lea, which seems quite distinct from those of the St. Johns River drainage. There were also fine specimens of *Ancylus peninsulae* Pils. & Johns.—erroneously referred to *A. filosus* in my list. The east coast canal has drained much of this section now called Bikes Prairie on the coast survey chart.

These notes suffice to show some of the changes that can take place in a limited area in a comparatively short time, and the probable effect of such changes on certain species. It is not at all peculiar to St. Augustine, for similar changes are going on at many other places along the coast and in the vicinity of our cities. The importance of a careful study of a local fauna cannot be too strongly urged. The destruction of the forests, the draining and filling of swamps and marshes, the construction of dams, etc., all tend toward lessening the fauna and flora of a given area.

THE MARINE SHELLS OF SANIBEL, FLORIDA

BY WILLIAM J. CLENCH

William James Clench, now Professor Emeritus of Harvard University, was born October 24, 1897 in Brooklyn, N. Y., but was brought up at his parents' home in Dorchester, Mass. His interest in insects and shells was crystallized in the direction of higher education by C. W. Johnson, then Curator at the Boston Society of Natural History. Clench received his doctorate degrees from Michigan State and the University of Michigan. During this period he was influenced by Bryant Walker and Calvin Goodrich, both eminent freshwater men. Clench went on to Harvard where he built up their collections and produced a host of outstanding students.

Late in the summer of 1921, a trip was undertaken to Sanibel Island, Florida, for the purpose of collecting mollusks. Arriving there the second week in August, collecting was carried on until the middle of September.

This cresent-shaped island is located on the west coast of Florida, about 130 miles north of Key West and two miles from the mouth of the Caloosahatchee River. It is sixteen miles long and five miles wide near the center. The eastern end of the island runs out to a narrow point, at the extremity of which is located the Sanibel light-house. The north-west portion is somewhat broader and is separated from Captiva Island to the north by a shallow strait known as Blind Pass.

Mollusks were found along the entire beach fronting on the Gulf and were especially abundant near the light-house, where the sand bars were much wider and longer. Collections were also made at Clam Bayou, a small shallow bay opening into Blind Pass, and at Tarpon Bay, a large bay on the north side of the island. Tarpon Bay, with the exception of a narrow outlet, is surrounded by a mangrove swamp, and all species listed as from Tarpon Bay include this swamp area, as well.

Many shells formerly common on Sanibel have disappeared, while others have become quite rare. The abundance of shells on the island, especially the larger and more showy species, attracts many tourist-collectors during the winter season, for the purpose of collecting. This might in part explain the paucity of many of these forms that were abundant a few years ago. It is quite probable that the shifting of many sand bars and the destruction of portions of the beach by wave action which has taken place in the last few years has also aided in decreasing the shell fauna. Examples of this are *Oliva sayana* and *Strombus pugilis*, at one time very common but now found but occasionally.

There were many species, however, which were very abundant along the entire beach, such as *Terebra dislocata* and *Donax variabilis*. The former was the most common of all the beach species and at low tide the exposed bars were covered with a network of their irregular tracks, the animals themselves being partially buried in the sand. The *Donax* was found in patches eight to ten feet wide, the location of which could easily be determined from a distance by the groups of shore birds feeding upon them. Dead specimens of *Cardium robustum* were found in abundance along high-water mark, and occasional living specimens on the outer sand bars.

In the vicinity of Tarpon Bay, *Litorina angulifera* was very common on the roots and lower branches of the mangrove. *Melampus bulloides* and *M. coffeus gundlachi* were collected in the thickets bordering the bay, and in the bay proper *Cerithium minimum* with the two varieties *nigrescens* and *septemstriatum* were usually found on exposed sand flats and in all shallow water, especially on submerged palmetto logs. *Busycon perversus* and *Melongena corona* occurred on both the beach and bay sides of the island, the beach forms being lightly marked and heavy, the bay forms usually covered with marine growths and of a lighter weight. Large clusters of the "coon" oyster, *Ostrea virginica*, covered the air roots of the mangrove trees and the wharf pilings. These "tree oysters" as they are sometimes called, presented a striking appearance at low tide when they were exposed to view.

THE NAUTILUS.

Vol. vi.   JULY, 1892.   No. 3.

SOME REMARKS ON NEW JERSEY COAST SHELLS.

BY JOHN FORD.*

Of the thousands of visitors to Atlantic City, Cape May and adjacent seaside towns, perhaps not one-tenth part give a thought to the myriads of living creatures other than human that sport in the surf, dally in the pools or hide in the sheltering sands. Yet it is not unlikely that the most superficial examination of these lowly forms would convince the observer that even seaside resorts may yield nobler pleasures than those of a physical nature only.

How many of these persons, I wonder, know that the despised Sea Nettles (Medusæ) often exhibit forms of surpassing beauty, rivaling in structure the most delicate of laces! And who of all the vast crowd think it worth while to note the wonderful variations in structure of the many species of crabs, shrimp, sandhoppers and other crustaceans dwelling between tides, and in some instances, in sands above the surf? Yet few if any phases of animal life, not even the transformation of a caterpillar to a butterfly, are as

---

* Editor of NAUTILUS,
Dear Sir :

The thought has occurred to me that many of the NAUTILUS readers would be interested, now and then, in articles less technical and scientific than those usually presented in its columns. In order to test the matter I take the liberty of offering for insertion the subjoined chat regarding New Jersey Coast Mollusks and a few of their neighbors.   Very truly,

JOHN FORD.

remarkable as the periodic metamorphoses of certain species belonging to this order. Near the water's edge, when the tide is low, many other interesting creatures may be seen, including the sea anemones with their parti-colored crowns of tentacles; and pretty plant-like forms (Corallines) whose chief representative on the New Jersey coast is the so-called fox or squirrel-tail *Sertularia argentea* Johnson. To most persons this appears to be an ordinary sea plant, but the careful student knows that in each of the tiny cells adorning the undried specimen, dwells one of the little architects and builders of the whole graceful structure.

None of these creatures, however, are more worthy of observation or study than are the native mollusks, reference to which is the chief purpose of this article. These dwell on the entire coast in countless numbers, but they are seldom exposed in quantity except by southeastern storms or gales which, striking the beach breast on, often tear up and carry large masses of sand with

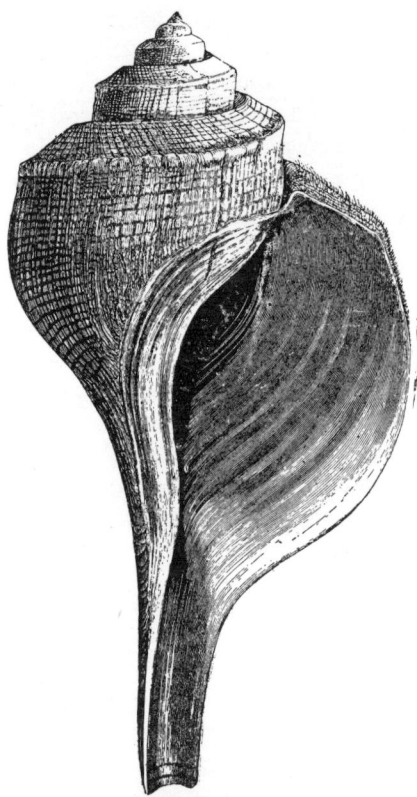

Fulgur canaliculata.

their unfortunate occupants beyond the reach of succeeding tides. It is not unusual for hundreds of tons of mollusks to be thus forced from their homes and left to die of starvation and exposure. Quite a number of the native species are edible. The first of these in the order of demand is, of course, the oyster, *Ostrea virginica;* next, the hard shell clam, *Venus mercenaria*; third, that precious favorite of all New York aldermen, the soft shell clam, *Mya arenaria.* He who has not eaten a dish of these on Coney Island beach would be deemed by the said New York magnates a "very unfortunate man" indeed. Less delicate in flavor than the latter species are the common sea clams, *Mactra solidissima*, when not more than

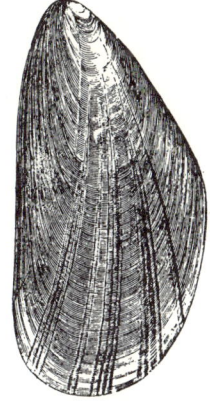
Mytilus edulis var. pellucidus.

half grown. Sea mussels, *Mytilus edulis*, are also considered palatable though they are not very highly relished in this vicinity. In New York and New England, however, they are much esteemed by epicures.

Among the fishermen of Long Island Sound the large Conch, *Fulgur carica*, is often utilized for soup. But the writer knows by experience that this is not the kind of food a delicate palate will long for.

Another edible species, and one far more toothsome, is the little periwinkle, *Litorina litorea*, a species probably introduced from Europe. Until recently they were quite rare south of Raritan Bay, but at present a fine colony may be seen on the flats a little west of the Inlet House at Atlantic City. The pretty species, *Litorina irrorata*, a more southern form, also edible, appeared in large numbers on the bay side, near Longport, N. J., about three years ago, but the conditions surrounding them changed shortly afterward and the colony disappeared quite as quickly and mysteriously as it came. A few specimens may still be secured on the adjacent flats but they are much less perfect than were those of the colony referred to.

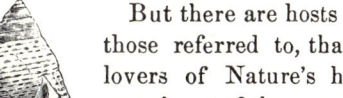
Litorina litorea.

Many of the shells produced by the several species mentioned are well worthy of a niche in the collector's cabinet, especially so if taken alive and *in situ*. Otherwise the more recent additions to the lip-edges are apt to be injured by the action of the surf.

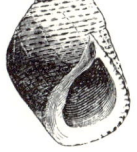
Litorina irrorata.

But there are hosts of other native shells besides those referred to, that may be profitably utilized by lovers of Nature's handiwork. Among the most prominent of these are the large pear-shaped Conch, *Fulgur canaliculata*; the several species of Pholades, including the largest known form, *Pholas costata*, which often secretes itself in the hardest limestone;

the canoe shells, *Modiola plicatula*; the razors, *Solen americanus* and *S. viridis*; the arks, *Arca pexata* and *A. transversa*; the boat shells, *Natica heros* and *N. duplicata*; the cup and saucer shells, *Crepidula plana*, *C. fornicata* and *C. glauca*; the ladder shells, *Scala humphreysii* and (rarely) *S. lineata*; the scallops, *Pecten irradians*, the adductor muscles of which are largely used for food, thousands of gallons being sold annually by the coast fishermen. In addition to these there are several small species belonging to the genera *Columbella*, *Nassa* and others, making the entire number living between Brigantine Inlet and Cape May about fifty species. At no special point, even on the most favorable occasions, can all of these be obtained. A large share, however, may at times be secured on the sea and bay shores near Longport, at Townsend's Inlet, Five Mile Beach and the Inlet two or three miles northeast of Cape May. But there is no locality known to the writer where species are so plentiful as at Anglesea; here, during a short visit last summer, thirty-nine species were secured by him.

Pholas (Zirphæa) crispata.

Scala Humphreysii.

Nearly all of these were found living on a small peninsula about a half mile south of the Anglesea Hotel. *Fulgur carica*, the largest of our coast shells, were unusu-

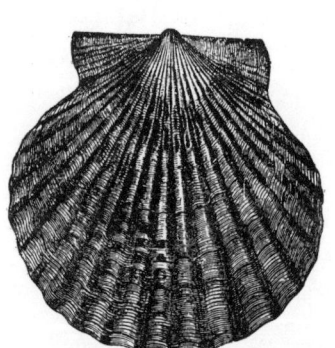

Pecten irradians.

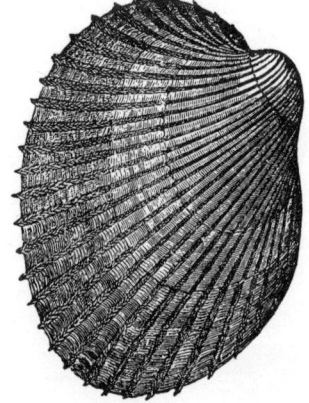

Arca pexata.

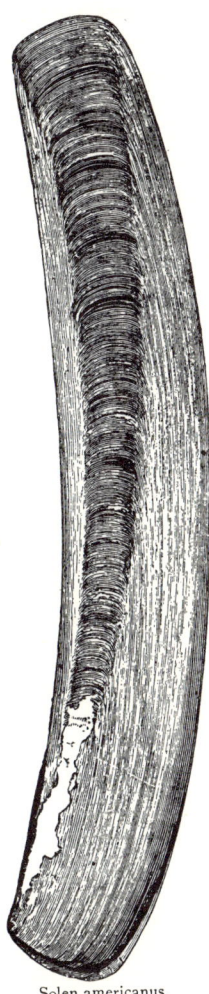

Solen americanus.

ally abundant, many of them being perfect in form, and exhibiting in the apertures the rarest shades of crimson, purple and orange. Excellent specimens of *Natica duplicata* were also found here *in situ*, these offering a new revelation to the collector as he saw, when lifting them from their beds, fine jets of water spouting in every direction from the edges of each large saucer-shaped foot. Both of these species were carried to the hotel and boiled—the former about ten minutes the latter two or three minutes. This made the removal of the animals an easy matter, leaving the lustre of the shells and color of the apertures uninjured. It should be remembered that the peninsula referred to is only free from water when the tide is nearing its lowest stage; also that the tide is low there at about the same hour it is high in Philadelphia.

In conclusion it may be well to inform the young collector that in the search for sea shells at least three adjuncts are necessary—a trowel for digging purposes, a water-tight jar for preserving living specimens and a good sized basket for large shells. With these in hand, and a taste for the work, there is no reason in the world why he shouldn't be both successful and happy.

SPECIAL NOTICE.

In order to increase the growing circulation of THE NAUTILUS in Foreign Countries, a limited number of copies will be sent to resident students in *Exchange* for desirable named shells. Address, THE NAUTILUS, Mt. Airy, Philadelphia, Pa.

NOTES ON MARINE MOLLUSCA ABOUT NEW YORK CITY.

BY ARTHUR JACOT.

Arthur Paul Jacot was a forestry expert trained at Cornell and Columbia Universities. He was born on Staten Island on March 28, 1890, and his first job was at the nearby American Museum of Natural History in New York City. Professor Jacot taught for ten years at Shantung University in China. From 1934 until 1939, the time of his untimely death, at the age of 49, he was with the U.S. Forest Service. He contributed eight articles to **The Nautilus,** *but also published elsewhere on ecology, fish and soil mites.*

Due to the unusually severe storms of the past winter the beaches about New York City were of special interest to the conchologist. On the one hand, bungalos, hotels, etc., were swept into the ocean bodily, while on the other, great quantities of shells were strewn along the shores.

At Rockaway Beach from the hospital (beyond the Park) to Edgmeer, a distance of four miles, there was an almost continuous rift of "skimmers" (*Spisula solidissima*) along the extreme high-tide line, which averaged two feet deep by ten feet wide. At some places these clams were piled up three to four feet deep, at other places they formed a double rift, while at still others (besides the rift at highest tide line) they were strewn as a thick carpet over that part of the beach laid bare at low tide. Counting 50 individuals per square foot, we estimated there were at least 5,000,000 per linear mile. It will be interesting to notice the abundance of this species along this strip of beach after next winter's storms. It seems as though a large colony has been whipped out for a distance of several miles. At Long Beach (the next beach eastward) there were a dozen or two *S. solidissima* per linear rod.

We tramped over the top of this gigantic funeral pile for a distance of half a mile, starting near the hospital. As a result we picked up 16 specimens of *Pecten magellanicus* (with the animal) besides a few odd valves. Beyond the half mile the species was not found. The most astonishing find of the day, however, was a large specimen of *Buccinum undatum* fairly well covered with *Hydractinia echinata* and occupied by a putrid hermit. This species is of accidental occurrence west of the eastern end of the island. It is not an inhabitant of our sandy beaches.

On the channel side of Long Beach where dredges are widening the channel and building up land we found some hundred-odd specimens of *Ensis directus* freshly cast up. The largest had a width of $1\,^{5}/_{16}$ inches and a length of over 7 inches. A very few were found which were perfectly straight-looking from the outside like a true *Solen,* but when they were cleaned out and the hinge examined they were found to be *Ensis.* As I have never found the slightest indication of a Solen in this vicinity, nor as I do not know of a single authentic record of one for our coast, I am forced to the conclusion that the lots in the American Museum collection (as well as those in other museums) labeled New York and forming parts of the collections of shell fanciers have (as is the case with many other species) the wrong locality attached to them through carelessness or ignorance. This error seems to have been a common one for this species and it is time it be definitely rectified. This mistake of locality in collections of collectors of shells as a hobby (as the Jay, Haines, Newcomb, etc., collections) is common, so that their collections should not be considered for distributional records of species.

PURPURA LAPILLUS L.

(Extract from the report of Mrs. D. J. Wentworth. From the Transactions of the Isaac Lea Chapter for 1896.)

No shell is more common on our New England shores than *Purpura lapillus* Linné, and yet, no shell of this region has to me, at least, so much of interest connected with it.

Purpura lapillus is an humble but most worthy descendant of the aristocratic Muricidæ, and surely the Murex is an aristocrat among shells, with its beautiful forms, dainty sculpturing, delicate coloring, and its long traditions of usefulness and importance.

Plain in its general aspect, as it certainly is, *Purpura lapillus* has nevertheless, much in common with its more highly favored relations. It is an old member of an ancient race, fossil remains of *Purpura lapillus* are found in the Red Crag deposits of Europe.

This species is remarkable for its variation in solidity of shell, form, sculpture size, coloring and habitat. It varies in thickness from three-sixteenths of an inch to a shell so thin one could easily perforate the outer lip with a pin. In form they vary from a short broad shell with obtuse spire and flattened whorls to a long shell with acute spire and convex whorls,

In some the coarse revolving ridges are barely discernable, while in others they are very prominent. The faint lines of growth which intersect the revolving ridges of this shell are, in some specimens, brought into marked prominence by rows of ruffles or scallops, and this sculpturing undoubtedly gave Lamarck reason for naming this variety *Purpura umbilicata*. Many of the solid shells have rows of nodules or teeth within the aperture on the outer lip,

These solid shells are usually grayish-white or white outside, with reddish-purple, yellow or white apertures; but the thinner shells are often brown, orange or lavender, and these colors are frequently banded with white. I have never seen two specimens banded just alike. The orange and white combinations are especially pretty. The variety called by Lamarck *P. umbilicata*, so far as I have observed, are always a greyish-white on the outside, with a reddish-purple aperture. This variety I have found only in a brackish river where they are often seen crawling about in the mud, and their color is so nearly the color of the mud on which they are found that it undoubtedly serves to protect them from the ravages of their enemy. Associated in this river with *Purpura lapillus*, and much resembling it in size, color and general shape, is the *Urosalpinx cinerea*.

Purpura lapillus is an arctic species and ranges from Norway to New York. It is found on the coast of Europe, where, according to Sowerby, it grows much larger than on our own coast. This species confines his daily rambles to that part of the shores left bare by the tides, seldom venturing below low water mark. There on the rocks or other hard substances he finds his favorite food, the succulent barnacle, sometimes varying his diet with a choice bit of *Mytilus edulis*, to obtain which he will bore through the shell. Finally the mussel becomes so weakened that its valves fly open, when the Purpura promptly accepts the more favorable opening and proceeds to gorge himself with the delicious morsel, after which he will lie inactive waiting for a return of appetite. Limpets, Littorinas, clams, mussels, etc., are said to find a place on his menu.

From time to time throughout the year the Purpura deposits its eggs enclosed in little vase like capsules. These capsules may be found in clusters attached to the undersides of rocks. In confinement it takes about four months for these eggs to mature and then the young do not immediately leave the capsule, seemingly preferring to try their strength a little before venturing on the broad ocean. The young hatched in captivity instinctively leave the water every day, remaining out about the time it takes for the tide to ebb and flow.

A few years ago while fishing I had occasion to crack some Purpura for bait. After cracking their shells I placed the snails in my handkerchief to keep them safely until needed. I soon found that the snails had stained my handkerchief with bright purple spots which repeated washings only served to render more brilliant. Thus I was reminded of the Tyrian purple of the ancients, and led to fancy that perhaps in a somewhat similar manner, the dye was discovered. In later years this dye was manufactured in Ireland but so little was obtained from each animal, and other cheaper dyes being discovered, our humble shell-fish were left to die a natural death, and are now useful only to amateur fishermen and so-called " queer people," or " cranks " who go around collecting shells and studying them.

Purpura lapillus commonly called " dog winkle " by the English, has many scientific names, among which are *Buccinum lapillus*, *Tritonium lapillus*, etc. But what is a name? The *Purpura lapillus* under whatever name he has crawled or sailed has a long, interesting honorable history,

SHELL COLLECTING AT MT. DESERT, MAINE.

BY JOHN B. HENDERSON, JR.

The coast of Maine has been thoroughly explored by biologists for many years, and has, indeed, become a classic ground in the annals of American conchology. Frenchman's Bay and the waters immediately about Mt. Desert seem to have been less exploited than other localities in Maine. Collectors of marine invertebrates going " down East" generally take their dredges and trawls to Casco Bay, or, if more ambitious, they hurry on to the famous old collecting region about Eastport and Grand Menan. A few notes from the shores of Mt. Desert Island may, however, prove acceptable.

Frenchman's Bay is a large body of water with a wide pass out to sea which is somewhat obstructed with bold, rocky islands. Through the openings between these islands the twelve and fourteen feet tides flow with great swiftness, scouring out the channels to a depth of from forty to fifty fathoms. In these deep places a tough form of algae clings tenaciously to the rocky bottom, and harbors within its tangle of branches and stems a vast multitude of small crustaceans (often phosphorescent), many curious star-fishes, and a wealth of molluscan life. *Margarita cinerea*, an occasional *Scala groenlandica*, abundant *Trophon clathratus*, *Bela turricula* and *decussata*, *Cemoria noachina*, young *Sipho*, and the lively little *Nassa trivittata* were observed. Dredging in these deep, rocky places is attended with many difficulties, but often yields satisfactory results.

The general average depth of the bay is twenty to thirty fathoms. The bottom is mud, with patches here and there of hard, pebbly ground, becoming rocky. These stretches of hard bottom are often the resort of great numbers of *Pecten magellanicus*, known to the natives as "scallops." This giant among the Pectens is gathered somewhat extensively for the markets, but does not make a particularly dainty dish. It is best collected by sinking or draging along a fishing-line over the bottom of the scallop beds. The big fellows seize the line viciously and permit themselves to be hauled out of the water; unfortunately, adult specimens are usually badly eroded.

Such stations contain *Crenella glandula*; they swarm with *Nassa trivittata*, and seem literally to be paved with *Nucula proxima*. The mud bottom is fairly rich in *Lunatia triseriata*, *Yoldia limatula* and *thraciæformis*, and again *Nucula proxima*. *Leda tenuisulcata* is occasionally met.

Passing out to the open sea the water very gradually deepens, and patches of shelly bottom are frequent. These places, made up for the most part of broken shells, fine gravel and sand, offer good rewards to the collector. *Dentalium entalis*, *Turritella erosa*, *Pecten islandicus* (dead), *Cardium pinnulatum*, *Astarte sulcata* and *Terebratulina septemtrionalis*, the latter, invariably imbedded in sponges, may be readily obtained.

Upon the rocks between tides, the usual Litorinas, together with *Purpura lapillus*, are always abundant, a splendid red variety of the latter occurring near Otter Cliffs. Just below the low-tide mark, *Chrysodomus decemcostatus* and a degenerate form of *Buccinum undatum*, range. Their home among the rocks protects them from the dredge, but they may be easily tempted by bait. In all rocky places of moderate depth the pretty little *Margarita undulata*, tinged with red and iridescent within, can be found.

On flats, exposed by the receding tide, of which there are a few in the vicinity of Mt. Desert, the soft clam, *Mya arenaria*, lives buried several inches below the surface. The number of these creatures annually taken by fishermen for bait from the "Bar" at Bar Harbor, figures well into the hundreds of thousands, yet the supply never seems to diminish.

A few dead valves of *Arctica islandica* indicates the presence of this boreal species in the bay. A more thorough examination of the depths of the harbor would undoubtedly reveal many more interesting things to the explorer than I came across in my two or three moderately successful dredging expeditions at Bar Harbor last summer.

Shells, Marine Curios, &c. I am now ready to supply first-class stock at low prices and should you wish anything from this section, let me hear from you. All inquiries will have a prompt reply.

J. H. HOLMES, DUNEDIN, FLA.

SHELL COLLECTING AT EASTPORT.

EDWARD W. ROPER.

Roper, who was born in Revere, Mass., October 12, 1858, was brought up on the farm of his uncle and aunt in Lynnfield, Mass. They moved to the seaside town of Revere, near Boston, in 1873, where he started collecting birds, wildflowers and shells. After graduating from the Chelsea High School, he held various editing positions with the Revere Journal, *Somerville's* The Truth *and the* Chelsea Record. *He married Miss Flora G. Allison, of Dublin, New Hampshire, in October, 1894. Roper suffered from continuing attacks of "Grippe", which could have been undiagnosed tuberculosis, and he finally left New England to escape the rigors of winter. He visited Jamaica in 1893 and 1894, then headed west, with his wife, remaining in Colorado Springs long enough for their daughter to be born. The family finally reached Pasadena in 1896, Long Beach in 1897 and San Diego in 1898. His main interest was in the freshwater pea clams (Pisidiidae) of the Pacific Coast, although he collected marine shells extensively both in Maine and California. Ill-health dogged him until the end when he died at the age of 40 in San Diego, on Dec. 31, 1898. His collection of about 3,000 species was left to the Boston Society of Natural History, and then transferred in 1930 to the Museum of Comparative Zoology at Harvard College. Pilsbry in 1889 named a* Triodopsis *land snail (roperi) that Roper collected at Redding, Shasta County, on his first visit to California. W. H. Dall named* Roperia roperi *in his honor, a marine species now known as* Ocenebra poulsoni *(Carpenter, 1864), also from California.*

The August number of the NAUTILUS was awaiting me on my return from a collecting trip to Eastport, Maine, with Messrs. B. H. Van Vleck and R. T. Jackson, of Boston, and I could fully appreciate Mr. Simpson's excellent article on dredging at Tampa Bay. Eastport is likewise "classic ground" to naturalists, and seldom a year passes that boatman Jerry Sullivan does not have an opportunity to take some ardent collector in his trim sloop. "Uncle" Jerry has been a resident of Eastport over forty years, and has coiled the dredge rope for Agassiz, Verrill, Fewkes and other well-known scientists. He knows the fluctuations of the strong tides, the depth of water, and what is of most consequence, the character of the bottom, which enables him to keep away from rocks which might cause the loss of the dredge.

While not equal to subtropical Florida as a collecting ground, Eastport, for a northern locality, is rich in species and individuals. Our dredgings were in water from fourteen to eighteen fathoms deep, and Mr. Simpson's statement that it was "hard, heavy, wet work," was certainly not overdrawn. Sometimes the dredge came up full of stones and gravel, with which were huge starfishes ten inches across the rays, curious leathery Boltenias, large red shrimps, sponges, such beautiful shells as *Trochus occidentalis*, *Margarita undulata* and *Admete viridula*, and perhaps the long-named brachiopod, *Terebratulina septentrionalis*. The best brachiopod ground, however, has been ruined, by the dumping upon it of blue clay dredged from Luber Narrows.

The best hauls were made on a moderately soft bottom of mingled mud and sand, which was literally filled with dead and living shells of *Cyclocardia borealis*, *Astarte undata*, *Astarte crebricostata*, *Cardium pinnulatum*, *Sipho pygmæus*, *Dentalium striolatum* and many others. Here also were obtained numerous brittle stars, *Ophiopholis*, and the *Astrophyton Agassizii*, which came up clinging to the outside of the net, nearly as often as inside. When the dredge landed in soft mud it brought up such shells as *Leda tenuisulcata*, *Nucula tenuis*, *Crenella glandula*, *Yoldia sapotilla* and *Cryptodon Gouldii*.

Shore collecting at Eastport is sure to prove successful. Ordinary tides rise and fall eighteen feet, and at low tide a large area of shore is uncovered. *Purpura lapillus*, *Acmæa testudinalis* and the various *Littorinas*, common all along the New England shore, are here of much larger size than in Massachusetts. *Buccinum undatum* is everywhere seen at low water mark, and bunches of its yellow egg cases are fastened to the rocks in abundance. Underneath stones are myraids of crawling things not well known to a conchologist, but nevertheless interesting. In the larger rock pools every stone hides specimens of *Chiton marmoreus* and *Chiton albus, Saxicava rugosa* and *Margarita helicina* are common and the bottom may fairly bristle with the spiny sea urchins.

The enthusiastic collector will understand my pleasure when a critical examination of my gathered treasures revealed about seventy-five species of shells, fifteen of which had not previously been represented in my cabinet. My companions, more interested in other invertebrate forms, were also quite successful. Add to this, the fact that we were in the coolest place in the country, wearing light overcoats many evenings while everybody at home was sweltering in torrid heat, and we may look back to our Eastport trip as favored by fortune and replete with pleasure.

THE ABUNDANCE OF CREPIDULA FORNICATA L. AT NANTUCKET, MASS. In all my collecting I never saw *Crepidula fornicata* in such great numbers as on July 14 along the north shore of Nantucket, from the bathing beach west for over a mile. A northeast wind was blowing and at high-water mark there was a windrow of shells from one to two feet wide. About one half of the shells were alive, attached in numbers to one another or in some cases to pebbles, and here also the shells were attached to one another. To the number of 7 or 8 the two clusters extending from the opposite sides of the pebble, the lower shell of each cluster being so closely fitted to the small pebble that the latter was often scarcely visible. This species was practically the only shell on the beach at the time.—C. W. JOHNSON.

A LARGE DECAPOD.—I have been greatly interested in an immense Cephalopod which came ashore about five miles south of Jack Mound, Anastatia Island. Only the stumps of the tentacles were left, as it had been dead for, perhaps, days. The body proper measured 18 feet in length, 11 feet in breadth and 3½ feet thick above the sand as it lay soft and flattened on the beach. Of course there is no way of knowing how long the tentacles were, but, judging from the size of the body, the arms must have been of enormous length.—DEWITT WEBB, M.D., St. Augustine, Fla.

CASCO BAY.

BY REV. HENRY W. WINKLEY.

The two most famous collecting grounds on the coast of Maine are Eastport and Casco Bay. The writer having spent several summers at Eastport, devoted his energies this year to Casco Bay. From the city of Portland to Cape Small the distance is perhaps thirteen miles. From the mainland to the outer islands is some six miles. This area is said to contain 365 islands. A fortunate location was secured on one of the outer islands, in a central position as regards the longer axis of the Bay. The naturalists of the expedition were the writer and his two enthusiastic and constant companions Frank H. and Robert L. Winkley aged 10 and 7½ respectively. The shores are for the most part rocky, affording occasional tide pools rich in animal life. The bottom is of every variety, giving opportunity for any taste the mollusca may display. Land shells abound on the outer islands. Singularly they find a favorite home here while on the main land they are exceedingly scarce. We visited, for land shells, Eagle, Brown Cow, Jewells, inner and outer Green and Cliff Islands; on all but outer Green we obtained good results. The most curious of this group is the famous Brown Cow. In the midst of rough ledges,—an out post fronting the open sea,—this mere spot, rises with perpendicular cliffs to a height of at least fifteen feet. The approach must be made in calm weather, and at low tide. We had a half hour's visit and such a harvest! The top of the island is one half covered with grass, the other half is a clump of bushes. *Helix hortensis* covered the leaves and branches of these bushes, the varieties being the yellow and five banded. On the ground *Pyramidula alternata, Polygyra albolabris* and *Succinea obliqua* were abundant. We obtained the famous wine colored variety of *P. albolabris*, and among the specimens discovered a set banded with fine lines, like *P. multilineata*. Time was precious and we collected expeditiously as the tide was coming in. We escaped from the island with a slight ducking from the surf, but happy are the results. On Green

island a few specimens of *H. hortensis* were found, among them
two full grown forms, which had for some reason started to grow
again; extending from the finished lip was a continuation of the outer
whorl, but of a dirty cream color and rough with ridges. On one
of the islands Frank discovered the home of the albino *P. alternata*,
a valuable prize. Shore collecting gave us a beautiful series of the
various varieties of *Purpura lapillus*, and some of the specimens were
the largest we have seen. We also found *Buccinum, Skenea planorbis,
Turtonia minuta, Rissoa aculeus, Lacuna vincta,* and the common
shore varieties. Considerable time was given to dredging in depths
from seven to twenty-five fathoms. One summer is far too short to
exhaust this region, but many localities were dredged with good
results. A dozen to fifteen new forms were added to the cabinet,
and at least fifty duplicate sets, to represent the Bay, found places in
the collection. Five species of chitons were found, including *Amicula
Emersonii;* a few fine specimens of *Pecten magellanicus* were dredged,
among them one that had received an injury and in repairing had
turned the edges of both valves upward so that they grew at right
angles to the natural plane. The interesting genus Bela revealed a
half dozen or more species, *harpularia* being the most abundant.
Brachiopods were found occasionally, and sponges, shrimp, echino-
derms and other invertebrates were abundant, but with much regret
at not having the means to care for them they were returned to the
sea. A list of results would contain all of the common forms. The
more rare species included the genera Thracia, Astarte, Nucula,
Modiolaria, Crenella, Cylichna, Margarita, Odostomia, Lunatia,
Velutina, Astyris and others.

Since the above article was written I have read with much interest
the article on "*Helix alternata*" by Mr. Ormsby. I do not wish to
take anything from his statements, but to add one or two concerning
that species. The islands of Casco Bay are good to stand a man on
his head, figuratively if not literally, for he meets with circumstances
which upset his former ideas. Land shells are very scarce in the
state of Maine, at least in the parts I have visited. As a rule two
or three specimens of the larger species, would be all one would find
after a careful search, not so, however, on the small islands. *Pyramidula alternata* occurs in great profusion. *Polygyra albolabris* and
Helix hortensis are also abundant. *P. alternata* occurs on one island,
some distance from any trees, just above high water mark, its only
shelter being rocks and small raspberry bushes. In this location
some two hundred, including the albino, were found.

COLLECTING MOLLUSKS ON A BEAM-TRAWLER.

BY A. B. FULLER.

Collecting mollusks on a beam-trawler while not an ideal way to collect, is nevertheless interesting. These vessels are about 135 feet long, 22 feet beam, tonnage about 150, and run by steam. August 6, 1920, found me on one of these steel boats bound for the Georges Bank. Our first set was made about 118 miles southeast of the Boston light in about 45 fathoms, inside of the Georges Bank proper. The trawl consists of a sweep net about 90 feet wide and 9 feet deep, held apart and in position by two heavy oak doors about three feet by seven feet, shod with heavy iron on one long side. This makes it ride upright and prevents it from wearing as it drags on the bottom. These doors act as kites to the net, as it were, one at each end of the opening, and each hung by a chain bridle to a steel cable. The cables are attached to steam winches which work simultaneously in lowering and pulling in the net. A heavy rope cable about three inches in diameter stretches from door to door and drags on the bottom, acting as a ground line to which the lower edge of the net is fastened. In the center of the net is a large pocket of coarse meshes, but smaller than the meshes of the net proper. This is called the "cod end," and is protected by a blanket of heavy double-meshed netting, so that in dragging on the bottom it will not snag and tear. A portion of the "cod end" is pursed and tied with a special knot; when the bag is hoisted aboard full of fish a pull on the knot opens the purse and the fish are dumped upon the deck.

The net is "fished" two hours at a time, and the time consumed in hauling, dumping and resetting is very short. The fish are then cleaned and are often all on the ice in the hold in about thirty minutes after being taken from the water. Three to four thousand pounds of good fish at a haul was fair fishing and about the average. Most of the fishing is in water ranging from 30 to 50 fathoms, in a zigzag course across the grounds.

The net sweeps nearly everything before it of any size and all goes back into the "cod end." The collection that is dumped upon the deck is therefore miscellaneous in character. From two to three tons of mixed fish, sponges, mollusks and other invertebrates is quite a sight to a collector. Large monk fish, skates, cod, haddock, hake, red snapper, halibut, flounders and sculpins, comprise the principal fish. Owing to the large mesh the majority of the mollusks pass through, leaving only the very large ones or a few of the smaller ones entangled in the net.

Each haul presented three chances to collect. First, when the net comes up; a few minutes of hasty inspection brought to light some fine nudibranchs (*Dendronotus frondosus*) and many little hermit crabs bearing various species of shells and a few very minute shells (*Cingula carinata*) imbedded in the strands of the ground line. Second, the fish are sorted by sluicing them down the deck with a stream of water, the men pushing the refuse fish along with pitch forks and picking out the good ones as this procession goes by, the shells, etc., may be snatched up and not much passes by without being seen. *Pecten magellanicus* Gmel. were sometimes very common, at other times missing. *Cyprina islandica* Linn., *Modiolus modiolus* Linn., *Buccinum undatum* Linn., *Chrysodomus decemcostatus* Say, *Colus stimpsoni* Mörch., *Polinices heros* Say, and the rare *P. levicula* Verr., were taken in this way, and all varied greatly in numbers according to bottom conditions. Some of the *Buccinum* and *Chrysodomus* were unusually large. Attached to some of the *Pecten* were the egg-capsules of *Chrysodomus decemcostatus*, called by the fishermen "sea corn." These were described and figured by Mr. Charles W. Johnson, "Occasional Papers," Vol. 5, pp. 1–4, pl. 1, 1921, Boston Society of Natural History. I am indebted to the Society for the cut illustrating these capsules. The third method of collecting is from the fish stomachs as the men were cleaning the fish; I was often able to get a bucket full of material from the haddock, later washing and sifting out the shells, wrapping them in cheese cloth and throwing them into a can of formaline. Sometimes the contents consisted mostly of small crustacea mixed with sand, with but few shells. The cod produced but little in the mollusk line except fragments of *Cyprina* and *Modiolus*, which they had evidently been able to crush. There were also pieces of large gasteropods, probably *Buccinum* and *Chrysodomus*. Crabs, however, seemed to be the main food of the cod. Sometimes the net would come up plastered with large starfish, then it would be a yellow sponge (*Desmacidon palmata*) that the fishermen call "boxing gloves," from their resemblance; another haul would show large numbers of ascidians, the "sea lemons," or the "stemmed sea peaches" (*Pyura*). Many times the net was filled with hydroids, known to the fishermen as "moss," clusters of long rubbery worm-tubes, dubbed by the men "macaroni," as it resembles that product, was very plentiful in one place. Thus the men would say, "we are on the boxing gloves," or on the moss, or in the lemons, or in the macaroni, as the case might be.

THE NAUTILUS.

ADVERTISING RATES.

Advertisements will be inserted at the rate of $1.00 per inch for each insertion in advance. Smaller space in proportion. A discount of 25 per cent. will be made on insertions of six months or longer.

NOTES.

CAUGHT IN A LIVING TRAP.—In the window of a Salem, Mass., store may be seen a unique sight, that of a kingfisher held tightly in the grip of a mussel. The story is this:

This forenoon patrolman Michael J. Little while crossing Beverly bridge, saw the bird fluttering on the flats, and he asked a fisherman to investigate. The latter went to the spot and there found the bird drowned.

It had swooped down and poked its bill into the open shell of a mussel, which suddenly closed on the bill of the bird. There the the two remained, until the incoming tide drowned the bird. Hundreds have viewed the singular sight today.—(*Boston Globe*).

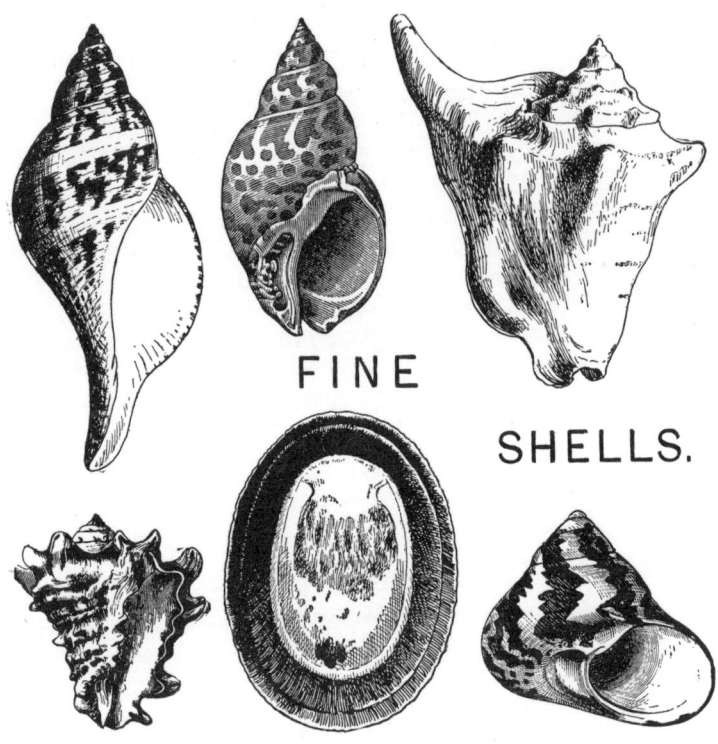

Send **40c.** FOR ILLUSTRATED DESCRIPTIVE CATALOGUE OF MOLLUSCA,

New Edition, 170 pages, over 200 Illustrations.

Just Received :—A shipment of fine shells of *Argonauta argo* attaining the remarkable size of 8 to 10¾ inches. Also several fine specimens of *Scalaria pretiosa*. Write for prices; also for list of *American Unionidæ*.

Ward's Natural Science Establishment

Pioneering on the Pacific Coast

Under the aegis of Josiah Keep, a professor of natural sciences at Mills College, California, for 26 years, a great cadre of enthusiastic amateur conchologists was assembled during the late 1880's. His excellent guide book, **West Coast Shells,** *with its various editions appearing from 1885 to 1911, was the only helpful book for beginners. A club was organized under the name of the Isaac Lea (a Philadelphia malacologist 1792-1886) Conchological Chapter of the Agassiz (Harvard's marine zoologist, 1807-1873) Association. Each member of the chapter was required to add an account of collecting activities to the "traveling transactions" that were passed among members. They had no publication outlet so editor Pilsbry published their interesting accounts in* The Nautilus *for several years.*

SUMMER STUDIES IN CONCHOLOGY.

BY PROF. JOSIAH KEEP.

Josiah Keep was born in Paxton, Massachusetts, in 1849 and received his B.S. and M.S. degrees from Amherst College in 1874 and 1875, respectively. He moved to Oakland, California, with his wife, Amelia, at the age of 29. He was first a principal of the high school in Alameda from 1881 to 1885, then later went to Mills College. He published 14 popular articles in The Nautilus, and was described by his surviving peers as "modest, courteous, indefatigable and enthusiastic as a teacher and organizer." Several of the following articles are by his students. Keep's article written in 1889, explains why he had so many enthusiastic followers. He died at Pacific Grove, within the sound of the sea that he so dearly loved, on July 27, 1911, at the age of 62.

For several years past a class in Conchology has been connected with the Chautauqua Assembly at Pacific Grove, Monterey, Cal. This Assembly meets annually about July 1, and continues its sessions for the space of two weeks. During this time there are numerous lectures, concerts, and other intellectual exercises, many of which are of a high grade of excellence. Such a programme, given at this delightful watering place, naturally attracts many visitors beside those who are engaged in the regular Chautauqua course of studies. The past season has been no exception, but the interest has been deeper and the attendance larger than on any previous occasion.

The science classes were mostly held at nine o'clock in the morning, and were followed by a public lecture. Excursions to the beach were made at various times, particularly in the early morning, in order to take advantage of the very low tides which occur then, about the time of the new and the full moon.

The class in Conchology was no respecter of persons in regard to age or occupation. Around the tables on which our shells were spread were seated matrons with gray hair, boys and girls, young men and women, ministers of the gospel, teachers from our schools, here a young man from the farm, and beside him a mother leaving for a little the duties of home. Perhaps in all the country, a similar class with a similar object could not be found.

And that object was the study of mollusks, particularly those species which were to be found in the immediate vicinity. Not so much a critical examination and discussion of the fine points of difference between similar species, but first of all a study of the structure and nature of the soft parts of the animal, then the mode of growth of the shell, the names of its parts, and its general morphology. After this, as far as time permitted, a study of the local species, and of others which have their home on adjacent parts of the coast. The apparatus was of the simplest kind. One morning a quantity of limpets were brought in for examination. Some were put into jars of sea-water and their motions observed. Others were deprived of life by a fresh-water bath, and distributed around the tables. After an examination of the foot, mantle, head, etc., penknives were used to slit the head, and common pins were employed to dissect out the buccal mass. A microscope was at hand to show the sharp teeth, and many were the expressions of surprise and interest in connection with the whole lesson. Valuable suggestions were made by members of the class, and many cabinets of shells were begun or received additions.

After a start had been made, the writer's book, "West Coast Shells," was freely used. Descriptions were read, engravings examined, and pronunciations recommended On the tables were spread numerous examples of dry shells, many of them imperfect, just as they had been gathered from the beach. From these mixed piles the members of the class drew out specimens of the shell under consideration at any particular time, and were given such hints and directions as would tend to fix its main features in the mind, and guard them on the one hand from confounding it with similar species, and on the other from separating it from its brethren on account of mere varietal differences.

The two weeks of study passed all too quickly, but even in that brief time good results were obtained. Not results of a critical nature, not important contributions to the science ; very few have the ability or the opportunity to accomplish these. But our science ought not to be simply for the scholar and the specialist. The people in the common walks of life have a keen sense of the beautiful, and the interesting features of common objects have but to be pointed out to be appreciated.

WEST COAST SHELLS.
By Prof. Josiah Keep, Mills College, California.

A familiar description of the Marine, Fresh-Water and Land Mollusks, found in the United States, West of the Rocky Mountains. 182 Illustrations, 230 pages. The frontispiece is a hand-painted engraving of the very rare and beautiful *Surcula Carpenteriana*, Gabb.

To aid in the study of Shells, the pronunciation of the Latin names is indicated, and a Glossary, Key, Biographical Index of Naturalists, and a Check List are added. Beautifully printed and bound. Price, postpaid, $1.75.

Helix (Aglaia) fidelis, Gray, Oregon. Samuel Carson & Co., Publishers, 208 Post St., San Francisco, Cal.

Please mention this paper.

Vol. XI.   AUGUST, 1897.   No. 4

SEEING EYES.

[From the report of Mrs. E. A. Lawrence. From the Transactions of the Isaac Lea Chapter for 1896].

Lowell says, "Eyes are not so common as people think, or poets would be plentier," and, what he has said of poets, could be said with equal truth of naturalists. Nature, to ninety-nine per cent. of the human family, is a closed book, not because she is not willing to have her pages opened, but because people have no eyes to see with. Thoreau could find in his back-door yard, or, on the shore of Waldon Pond, the material for printed volumes. It was not because these places had more of interest in them than similar places elsewhere, but it was because Thoreau had trained his eyes to see. How many people, of all those who yearly visit our sea-shore, have seeing eyes? They will tell us of the beauty of the foam-capped waves, or the brilliant tints of the water at sunset; but they will pass with unseeing eyes on careless feet over myriads of living creatures, creatures so wonderful in their mechanism, so beautiful in their form and coloring, and so cunning in their instincts, that they show to the observer the perfect workmanship of the great master, Nature.

About two years ago, I began to view the sea-side world through open eyes, and in that time I have collected over two hundred species of mollusca in the vicinity of San Pedro Bay. This past year I found beneath shelving rocks the little *Megetabennus bimaculatus* Dall, and nestling in hollows in the rough rocks I found a number of *Gadinia reticulata* Cpr.; the latter are so nearly the color of the rocks that it takes sharp eyes to discern them. Upon goose barnacles I found a small *Acmœa* which I hoped might be a new species, but I found my eyes were not very wide open even after a year's experience in the opening process, for I sent the *Acmœa* to Dr. Dall and he said they were only a small variety of the *A. pelta*. But in classifying my shells I have made two varieties of them, as I think there is enough difference in them to warrant such division.

Going down on the sand one day at low tide, I saw a small upheaval of the sand, and since I have been travelling the world with seeing eyes, I always investigate these tiny mounds, and, this time, I was rewarded by finding a *Tornatina culcitella*, and a diligent search soon revealed several of these cunning little creatures. With these are *T. inculta* and *T. carinata*. While out in a boat among the kelp I found a number of *Lacuna unifasciata* and some *L. porrecta*, these latter were three-eighths of an inch long. Among the tiny shells, seeing eyes are called into requisition, and I find the aid of a strong lens often necessary to bring out their distinctive features. The tiny *Marginella (Volvarina) varia* is more beautiful than a *Cyprœa spadicea*, and yet the blind world never finds them hidden away under stones, and covered with their thin tents, which are quite a protection for them, as it hides their shining surface and makes them much less conspicuous. The *Turbonillas*, the *Odostomias*, the *Mitromorphas* and many others have enriched my cabinet, and opened my eyes to the wonders and beauty of small things this past year.

When I began to classify, I had many a struggle with the different authorities, and many a dispairing appeal I sent to our General Secretary, who never failed to come to my relief, and I have much to thank her for, inasmuch as she has given me light where before I dwelt in darkness. * * * * To sum up the year's work, no greater gain has come to me than has come by the opening of my eyes, and the knowledge I have gained by seeing. The earth has taken on new beauty and the sea has opened some of its wonderful storehouse and bidden me enter, and all nature beckons me with kindly finger to further discoveries by the aid of seeing eyes.

COLLECTING CHITONS ON THE PACIFIC COAST.

Excerpts from a Diary.

BY MRS. M. BURTON WILLIAMSON.

Martha Burton Woodhouse Williamson was one of the leading lights among Keep's followers, and came to make some very creditable scientific contributions to the Proceedings of the U.S. National Museum and the Bulletin of the Southern California Academy of Science. She submitted 48 articles, many of a popular nature, to The Nautilus. Born in England on March 6, 1843, she came to America at about the age of 10, and was brought up in the Midwest where she later attended Burlington University in Iowa. She married Charles W. Williamson. Mrs. Williamson was a large and imposing woman, an active editor of the Terre Haute Enterprise, and a very ardent worker in the Womens Christian Temperance Union. She moved to Los Angeles, California, in 1887, and was active in civic affairs, including the Los Angeles Historical Society. She was secretary of the Isaac Lea Chapter, which was a shell group under the broader auspices of the then popular Agassiz Association. Mrs. Williamson died on March 18, 1922, at the age of 79, at Los Angeles.

Whilst peering under a rocky shelf (at Point Fermin) I saw something that seemed to move when I touched it accidentally with my knife. I pushed my knife under one end of it—the only end visible—and I found that the resistance was not that of a hard substance, but became less as my knife went farther under the rock and soon I had a big chiton, *Stenoradsia Magdalensis* Rve., on my knife! * * * Near where we found the *Conus* in the moss, in a shelving rock so close to another rock below it in the water that we could not remove it, we found a huge chiton. To get it out, it was necessary to break the sandy rock with a hatchet. There they lie, *Stenoradsia Magdalensis*, so close together that in less than three feet of the layer of rock that was chipped off we found over one dozen; some almost four inches in length ! Just as they were collected from their damp environment they presented a beautiful appearance. On the outside, as well as inside, save three or four old fellows, the shells were a bright pink, like the interior of a pink-lined sea shell.

In a mossy carpet on a wet rock I found chitons, *Mopalia ciliata* and *M. lignosa* imbedded in the rock. Sometimes the Chitons were entirely covered with moss and could only be detected because the moss seemed to be growing in a circle. In the moss the Chitons, on the outside, were green and brown like the moss around them; under and between the rocks, they were pink; when found in little depressions in boulders out of the water, they were almost the exact color of the stone on which they lived. These were mostly the *Chætopleura Hartwigii* Cpr. Here as elsewhere environment seems to play a most important part. Are specific differences merely the changing forms due to environment alone?

Vol. IX. JANUARY, 1896. No. 9

COLLECTING IN SOUTHERN CALIFORNIA.

[Report of Mrs. G. W. White. From the Transactions of Isaac Lea Conchological Chapter of the Agassiz Association for 1894.]

My interest in the science of conchology dates from the summer of 1893, when Prof. Josiah Keep, of Mills College, taught this subject during the Chautauqua Assembly at Long Beach, Cal. Though unfortunately, not a member of his class, some of my friends were enthusiastic students under him, and through them, the enthusiasm or " craze," as we somtimes call it, was communicated to me, and I very shortly became as eager a student and collector as my circumstances would allow. Happily for me, my husband shared my in-interest, and most of our collecting has been done together, and many of my choicest treasures were found by him in places I could not venture—in places where the waves dashed too high for my courage, or under rocks too heavy for my strength to lift. So this report must be understood as a record of our joint work.

In the winter of '93 and '94, we made a number of visits to Long Beach and San Pedro, going as far as Alamitos Bay in one direction and nearly to the old wharf beyond Times' Point at San Pedro. We collected quantities of shells, but as it would be impossible and unprofitable to mention them all, I will speak only of the rarer ones.

Under the rocks at Dead Man's Island, we found, on our first trip, a number of specimens of each of the following varieties: *Volvarina varia, Terebratella transversa*, and *Lazaria subquadrata*.

We also dug a number of fine fossil pectens out of the bank. One of the choicest shells found by us that day has never been fully identified by me. By some it is called *Cerostoma foliatum*, by others *Murex trialatus*.[1] At any rate it is a rare and interesting shell, and I have never seen another like it. On another trip up the beach above San Pedro we found, by digging in the sand with a trowel, some fine specimens of *Stenoradsia magdalenensis*, the largest Chiton on this part of the coast.

We also found several *Hinnites giganteus* Gray, *Cumingia californica* Conr., *Lucapina crenulata* Sby., and a fine old *Mitra maura* Swains., two inches in length, of which we are justly proud. In the spring of '94 we were on a visit in Ventura County, and when our friends proposed a trip to the beach, we interposed no objections. They drove to Punta Gorda, meaning Point of Rocks, most appropriately named, for I have seldom seen such a bed of rocks jutting out so boldly into the sea. They were literally covered with the largest species of mussels, many of them being nearly, if not, six inches in length. In the sheltered places in these rocks we found quantities of *Purpura saxicola* Val. and *Monoceros lapilloides* Conr. Our patient search in the rock pools was rewarded by our finding *Opalia crenatoides* Cpr.

In the summer of '94, while attending the class in conchology at Long Beach, our teacher, Mrs. M. Burton Williamson, kindly planned an excursion to Dead Man's Island, and took eighteen members of her class to San Pedro the evening before we were to do our collecting. We spent the night at an old seaman's hotel on Timm's Point, and at 3 o'clock A. M., rose to take advantage of the first beams of the sun and the tide, which was to be at the lowest point at about 4.30 A. M. There happened to be a dense fog, and as our ghostly boatman took one boat load after another of our companions away from us across the bay, we were strongly reminded of that other boatman, the Charon of our early studies and the River Styx.

However, by the time we were safely landed on our hunting grounds, the mist had risen and we could see to begin work. The most that we found of value was on the mud flats uncovered by the low tide. There we found, under the grass which lay flat on the mud, thus concealing thousand of mollusks which lay below, *Haminea virescens* Sby., and *Haminea vesicula* Gld. We also found, partly covered with mud, *Cardium quadrigenarium* Conr., and altogether covered with mud except some tiny points of a *Chorus belcheri* Hds. What a shout went around when some one called out, "Mrs. White has found a *Chorus*," and how eagerly the mud in that vicinity was scanned to see if another could not be discovered. But no, I bore off my trophy in triumph alone, for not another one was found.

On our way home, while walking along the beach, some one, I think Mrs. Williamson, called our attention to some narrow slits in the sand, where, upon digging carefully, we found a dozen *Lingula albida* Hds.

Later, on a walk to Alamitos Bay, I found *Periploma argenteria* Conr., *Petricola carditoides* Conr., *Labiosa undulata* Gld., *Yoldia cooperi* and *Clidiophora punctata* (?) Cpr. * * * * And now this account brings us up to the year of our Lord, 1895, and finds us still enthusiasts in conchology, only waiting for a favorable tide to go again in search of treasures of the sea. We are of those who believe that nature has secrets which she reveals only to those who love her, and we feel that in this kind of communion with her she has fully rewarded us.

LEAVES FROM A DIARY.

BY M. BURTON WILLIAMSON.

"We had thought the cliff at White's Point, Los Angeles County, hard to descend, but, when we saw the precipitous trail down which we were to pass at Point Fermin, we almost held our breath! After a while, C. and J. being in advance, they called to E. and I that the trail was not so bad after all. We slid down, or, jumped down, as loose dirt or stones were under our feet, and, sooner than we hoped for, we were on the rocky beach below. Almost at the top of the cliff I had found, in the sandy rock, the *Acmœa patina;* and the first shell I found on the wet rocks, was a live *Acmœa patina*, Esch.! On a great mossy bed of solid stone about 40 feet square I found the Conus Californicus, Hds. so thick, I was reminded of wild strawberry picking in my younger days. The *Conus* in almost every instance was partly hidden in the wet moss. Near this mossy carpet three *Cyprœa (Luponia) spadicea*, Gray, were found by C. and J. If the collecting of the *Conus* reminded me of picking strawberries, the *Luponia* in his shell with his red mantle dotted with bright yellow dots, was a huge strawberry himself! From under him rose his thin mantle until it almost covered his glossy shell. The shell shaded brown and drab, with a suggestion of the blue of the sky between the two colors, the transparent mantle, so gaily dotted with yellow, rising up over the brightly colored shell until it nearly met above in a frilled border, was a sight all five of us stood around and gazed at in wonder and admiration! Our delight found expression, then slowly the mantle was drawn down and out of sight."

MORNING TIDES.

[From the report of Mrs. M. L. Beck. From the Transactions of the Isaac Chapter for 1896].

One bright day in June I was told we were to go collecting the next day at San Pedro, and as the tide would be low at half past three in the morning, we would have to go to the beach the evening before and stay all night at the cabin on the island. * * We ate supper, and while two of us got things ready for the night, the rest went out on the breakwater to admire the scene. The high tide by moonlight was exquisitely beautiful.

At three o'clock, after having breakfasted, we started out to collect while the moon was still shining brightly on the water. The tide was so low it seemed to me we could have walked over to San Pedro. Mrs. O. and I lingered back of the other collectors, and soon she picked up a *Ranella californica* Hds., a fine specimen which now has a corner in my cabinet. How I did wish I could find one. I poked around with my trowel and suddenly I struck a lump; picking it up, it proved to be a perfect specimen of *Pleurotoma carpenteriana* Gabb, four and a quarter inches long. As I was afterward told, the only live one found in the bay. After returning to the cabin we put it in water, and when disturbed it exuded a purple fluid.

We walked to Dead Man's Island and found a number of *Actæon punctocœlatus* Cpr. in the pools, *Marginella Jewettii* Cpr., *Phasianella compta* Gld. clinging to the sea grass on the rocks; plenty of *Fissurella volcano*, *Chlorostoma aureotinctum* and *Littorina planaxis* all along the breakwater. On our way back to the cabin we collected *Haminea virescens* Sby., *Bulla nebulosa* Gld., *Conus californicus* Hds. and *Nassa tegula* Rve. We also brought home a good many Chione, from which we made delicious soup.

In July we went to Alamitos Bay, five miles from Long Beach; it was another fine low tide. This time seven of us went in a wagon at four o'clock in the morning. We found *Crucibulum spinosum* Sby. on oyster shells, *Cerithidia californica* and *Melampus olivaceus* crawling up the grass stalks near the edge of the water, *Œdalia subdiaphana*, *Angulus variegatus*, *Liocardium substriatum* and *Donax flexuosus* living as it seemed in harmony together, also *Amiantis callosa* Conr., *Tapes staminea* Conr., *Olivella bœtica* Cpr., and many other shells.

ODOR OF SNAILS.

It may not be known to every conchologist, that some of the Helices have odors peculiar to them.

We find here, *Mesodon ptychophorus*, *Patula strigosa*, *P. solitaria*, *Triodopsis mullanii* var. *olneyæ* in the same locality. The *Patula solitaria* has so strong an odor, like *Mephitis mephitica*, that I supposed at first they fed on *Ictodes* (*Symplocarpus*) *fœtidus*. Always the same odor and at all seasons.—MARY P. OLNEY.

MARINE SHELLS OF PUGET SOUND.

By Mrs. Marie Drake. From the Transactions of Isaac Lea Conchological Chapter of the Agassiz Association for 1894.

I have a *Glycimeris generosa* Gld. which I got from Dr. Pomeroy of Vashon Is. It weighed 7½ lbs. when alive, and was dug from a depth of three feet. Its length is 7½ inches, width 4¼ inches. Its longest circumference is 13¼ inches. It gapes widely at both ends; rarely meets when alive. Its edges are covered with a yellowish-brown epidermis. The pallial sinus, though not very deep, is from ¼ inch to ½ inch wide. Its distinct concentric grooves or lines are slightly irregular. The valves of this shell are strongly bulging. This shell is commonly called "Goe-duck," because it is so deep a burrower. The Indians esteem this shell-fish a great delicacy, and ornament their houses and yards with the shells. It is highly esteemed as an article of food, though quite difficult to obtain; one is said to furnish food for a whole family. Star-fish are also found by the hundreds at low tide on the mud flats, and of every hue—bright scarlet, peacock-blue, sea green and paler tints.

I have found three species of Purpura here and many varieties, but the handsomest one is *Purpura crispata* Chemn. In my mind, the finest shell belonging to Puget Sound is this Purpura when banded orange and white. This shell usually does not see the light of day. Some persons prefer the deep orange variety. Both live under the water on the under side of stones or in rocky places, and are either obtained by dredging or at very low tide. *Purpura lactuca* Esch. is not found in so deep water, hence its white color; it is exposed to the rays of the sun sometimes. About September 1st, you can notice a great many purpuras closely packed together, clinging to rocks laying their eggs, which are in little capsules and resemble yellow oats stuck on end thickly over parts of the rock. Each capsule contains three or four dozen eggs, which require about four months to hatch, if they are not doomed before by some starfish hungry for an egg dinner. You can find a few egg cases almost any time of the year, but most of the eggs are laid during the months of August and September. *Purpura* lives on mussels, limpets and barnacles, or, if food gets scarce, it will eat dead fish. But the Purpura are not always victorious, for, when a crab wants a "purple tea," he shows no mercy to the destroyer of other homes, but inserts his strong claws under the operculum of the Purpura and digs him out and devours him.

One June morning, at Point Defiance, I saw three *Calliostoma costatum* eating a sea-weed breakfast. They looked so dainty and seemed to enjoy the bright sunshine so thoroughly, that it was with some regret that I placed them in my basket.

One of my friends dug up a fine *Priene oregonensis* Redf. It was five inches long and covered with a heavy dark brown epidermis. When the epidermis is removed, the shell is white. It has a strong epidermis. I have found a few specimens of *Bittium filosum* Gld., under stones at low tide, and several *Margarita pupilla. Crepidula dorsata* Brod. I have found by the hundreds growing on the shells, especially upon *Placunanomia macroschisma* Desh. The Littorina are very plentiful and are large. I have searched for *Chrysodomus dirus* Rve., but have seen no traces of it. Perhaps it is found only on the ocean beach, and does not care for the Sound.

Modiola modiolus L. (*Modiolus capax* Conr. ?) grows to an enormous size in the vicinity of Puget Sound. My husband brought me several from Henderson Bay; the smallest measured 7 inches in length, the largest 9 inches, and was 4 inches in diameter. These he found growing in the mud, standing perpendicular, only about an inch being visible at very low tide. They are heavily bearded near the edge, partially covered with a light brown epidermis (which is several shades lighter than the epidermis of this same species which grows in the south), and considerably eroded near the unbones. All the shells living in the mud here are somewhat eroded. These monsters have an uncanny look; they are hermits when they grow old, do not live in clumps or groups as they do when young, or as *Mytilus edulis* does. Often I have seen a solitary *M. modiolus* upon a pile or log, which was entirely covered with *M. edulis*. They grow from $\frac{1}{18}$ inch to 8 inches in a single year. It takes muscle to remove one of the huge creatures from a rock or pile when it has fastened itself with brown byssal threads, which it spins with its huge tongue-like foot, from a sticky secretion formed at the base of the foot. They are said to live six hundred feet deep down in the ocean. Pupuras are death on mussels.

Placunamomia macroschisma Desh. is found here in great numbers. They live upon the under side of rocks which lie wholly in or part in water. A chisel is necessary to separate them from the rock, and even with this the pear-shaped byssal plug is rarely obtained entire. The interior of the upper valve is of a lovely sea-green and nacreous. The edges of the valves are thin and crumble at the least touch, which renders them difficult to clean and send away. If they grow upon other shells they are not so easily broken, but are much stronger. I have a fine specimen which I found growing in an old shell of *Cardium corbis* Mart. I obtained the byssal plug and both valves entire. The shells sometimes grow upon each other; when thus found, a perfect specimen is more readily obtained than from a rock. These shells are often mistaken for an oyster, especially by those unlearned in shell lore; they do resemble the variety known as *O. expansa*, though they are much larger and have the byssal opening and plug, which the oyster does not have. These bivalves are much handsomer than their southern cousins *Anomia lampe*. The animal is a bright orange, and is quite beautiful. To be prepared for the cabinet they are dipped in very hot water and the animal removed with a tiny steel chisel prepared for the purpose, then gently closed. This shell requires careful handling.

I saw a *Lunatia lewisii* Gld. eating a *Cardium corbis* very much larger than itself. I stopped this predatory proceeding, took both home in my basket, and, after cremating the bodies, placed both shells in my cabinet. Both these shells are abundant on the Sound, and are easily obtained by digging in the sand and mud. The *Macoma* family thrives here. I have not found *M. secta*, but *M. nasuta* and *M. inquinata* are prized by the Indians as food, and *M. inconspicua* is found by the hundred, the exterior slightly eroded by the mud in which it dwells, but the inside of the shell is of a bright, rich, shiny pink; pale yellow and pure white are also found. The shell is about the size of a finger-nail.

I was surprised to find upon the rocky beach at Brown's Point a living specimen of *Lyonsia californica* Conr. It was moving about in a pool of water among pebbles and rocks. It seems marvelous that its thin, delicate little shell could remain uncrushed an instant; but it seemed to enjoy life as well here in the rugged, stormy north as it does in the warmer waters of the " land of sunshine and blue skies."

Cardium corbis is more hardy, though, unlike most of our northern shells, it is smaller than its southern cousins *C. quadragenerium* and *Liocardium elatum*, but it is much more numerous than either of these species.²

Pecten hastatus Sby. is called by many our "prettiest shell," and with the thousands of little spears (*hastatus*), toothed edges and delicate coloring, it is indeed a lovely shell. I saw one for the first time (living) at Point Defiance during the month of April. It was caught on the top of a rock by becoming entangled by a piece of sea-weed; it opened and closed its shell rapidly, making a curious sound. The orange color of the animal shone and glittered in the sun. The circulation could be seen and the working of the heart and other organs. This Pecten is a deep-water species, swims about freely in the water and moves about at the same angle as a kite does in the air. It lives among sea-weeds and is found in great abundance at Fox Island in the spring of the year. The lower valve is bleached by the sand. It lives in the water and is never exposed to the rays of the sun, hence the delicacy of the color. This shell-fish has black eyes, and can tell when a hand or a bird comes to grasp it. I have seen *Amusium caurinum* in deep water, but have never succeeded in capturing one, as it is obtained only by dredging in very deep water. It is brown outside, white within, and has 20 ribs, and is not so handsome as *P. hastatus*.

Mya arenaria L. is highly prized here as food, and grows six inches long. *Machœra patula* Dixon is sold in our markets. *Psammobia rubroradiata* Nutt. is more abundant and larger than in the south. It is found 5 inches long, here, partially covered with dark brown epidermis. Tapes and Saxidomus are well represented, and, though not so prettily marked, are very much larger and stronger (coarser) than those growing in the warmer waters of the south. They are almost always to be found in the markets.

Zirphœa crispata L. was recently described in one of our Tacoma daily papers by one of our Government surveyors as " a new clam."

" We have found a new shell unknown to science," etc. We were greatly amused, and sent an article to the paper the next day saying *Zirphœa crispata* (" a new clam ") is found in abundance on both sides of the Atlantic, and was named by Linnæus *long* ago.

Limpets I have found in great abundance and of great size. I have several specimens of *Acmœa patina* Esch. found here in the " Narrows," measuring 2¼ inches in length and 1¾ inches across.

Many of this species have bands of translucent tints on their interior, and are beautifully marked outside.

A. pelta is regularly marked with stripes from the apex, which is often corroded, in adult specimens, to the base. This is a most pleasing shell; is a sort of hermit, lives alone, often easily obtained; strong, not easily broken; often pure white inside, sometimes banded. I have one with a bright yellow band inside, embossed. Large specimens measure 1½ inches in length, 1¼ inches wide, 1⅛ inches high. I have not found *A. spectrum* Nutt. nor *Lottia gigantea* Gray, here. Fine specimens and many variations of *A. scabra* Nutt. are abundant. I have found more of *A. persona* than of any other species. At Brown's point, we find at one spot a variety having a gray interior with beautiful translucent bands. This is a new variation to me. *A. asmi* Midd. is found here, and many I have not been able to classify.

I have one specimen of *Fissuridea aspera* Esch. 2¼ inches long, 1½ inches wide and nearly 1¼ inches high.

Limpets are sometimes used for picture frames by setting them deep in wood and fastening with glue. I saw one valued at fifty dollars here.

ON A COLLECTING TRIP TO MONTEREY BAY.

BY WILLIARD M. WOOD.

The editors of the NAUTILUS have asked me to write a short article for the NAUTILUS, while I am here, on my trip to this once famous collecting ground.

Now that I am about to leave for San Francisco, I feel sorry to think that I have not devoted more time to the collection of specimens. Of course, there have been many long drives to be taken, a dip in the surf once in a day, huckle-berry expeditions with friends, and a thousand and one things to be done, while stopping at a summer watering place.

Between these "sports," if I may be permitted to call them such, I have managed to find time to do some collecting.

The hotel at which I am stopping is situated within five hundred yards of the beach. To the north, runs a very smooth beach, devoid of rocks of any character for some fourteen miles. To the south, and extending for many miles, is a very rocky stretch. To this rocky portion, almost all of my collecting trips were confined.

Monterey is no longer the famous collecting ground it used to be. The increasing population at and around Pacific Grove is driving away all the land shells. The deadly sewerage flowing from the various towns' into Monterey Bay is killing the marine shells. However, new and very interesting species are occasionally brought up from deep water by the dredge.

Early in the morning, on the 28th of June, I started by steamer from San Francisco with my shell collecting outfit, consisting of glass pill bottles for small shells, paper boxes, cigar boxes, cloth bags, long, thin pieces of wood with rubber bands attached for the Chitons, alcohol stove and pan for the killing of bodies of the shells, cotton batting, long rubber boots, an immense sun hat, a chisel to detach Haliotis shells from the rocks, etc.

I arrived here at seven in the evening and although the trip down was rough, and our little "tub" rocked dreadfully, causing me to be sea-sick, it nevertheless did not prevent me from starting right in and collecting as soon as my feet rested on terra firma. On that evening, I began collecting at seven o'clock and as it was very light at that hour, I continued to collect along the beach until eight. I am very glad I did so, as it netted me some beach-washed species which I have not come across since.

I selected a week when the early morning small low tides occurred. Thus, one morning I devoted to the collection of Haliotis cracherodii, another morning I went in search of Littorina planaxis, another for Chlorostoma costatum, Acmæa scabra, Nassa mendica, etc.

During this second week, when no morning low tides have occurred, I have gone among the rocks, gathering any and every species which was so unfortunate, nay, I should say, fortunate, as to be placed within my reach.

Priene Oregonensis Redf. will be noted as having been collected here. I do not as yet understand how this large and beautiful northern shell should be found so far south. It could not have drifted into the bay, as it was a fresh, perfect-lipped specimen.

I may also mention that in a letter recently received from Mrs. M. Burton Williamson, of University P. O., Cal., that lady informed me that *Psammobia rubro-radiata* Nutt., is not found north of San Pedro Bay. As will be noted, I found one specimen, alive and perfect. It is truly a beautiful shell. The inside of both valves resembling delicate porcelain.

I am exceedingly sorry to think that I have no dredge here with me, as I feel positive I could gather at least five times as many specimens as I have already collected.

THE PURPLE DYE OF MOLLUSKS.—Any one desiring information on this subject can find a most interesting and comprehensive account in a paper entitled, "The Dyeing of Purple in Ancient Israel," by Rev. Isaac Herzog, published in the Proceedings of the Belfast Natural History and Philosophical Society, session 1919-1920, No. 2, pp. 21-33. As a summary the author says:—"The art of purple-dyeing in general, which dating from hoary antiquity—the mention of *tekelet* and *arganum* in the cuneiform texts occurs already about 1600 B. C.—passed through a long and checkered career, finally becoming extinct, at least in the Old World, on the fall of Constantinople 1453."—C. W. J.

Vol. VII.   NOVEMBER, 1893.   No. 7.

SAN PEDRO AS A COLLECTING GROUND.

Sarah P. Monks.

San Pedro, California, is remarkable for the number and variety of recent and fossil mollusks.

New forms and an unusual abundance of known species are constantly being found.

This is due in a great measure to the extension of the Government breakwater, which has made changes in the sea currents near the shore, and caused the tide water of the harbor to scour out the channel and drift large quantities of sand over the shallows.

By this means new homes are made for wanderers, and old inhabitants are washed from their moorings and swept by the tide within reach of eager Conchologists.

It is surprising, however, how seldom the year's abundance of any species repeat themselves.

At one time *Nassa fossata* Gld., at another *Periploma discus* Stearns; at another *Lima orientalis* Cpr.; or *Scalatella striata* Cpr., are found by the dozen, or score, or hundred in San Pedro Bay or vicinity, and then for years after only a few are found at a time.

The sea conditions are unsettled. This keeps local collectors alert.

Within a few months I have found a specimen of *Tritonium gibbosum* which is new to California, and one of *Cylichna cylindracea* var. *attonsa* Cpr., which is new to San Pedro. Both shells are beach worn.

This summer I spent July at San Pedro and added a number of new specimens to my collection besides learning many interesting facts about habits and habitat of molluscs.

A student only gets a half knowledge who cannot collect specimens and study the living animals in their native haunts.

July seems to be a favorite month for many species to lay their eggs.

Mitra maura (*Idæ*), fastens her capsules to the underside of stones; the Naticidæ place their "sand collars" in the damp sand; *Bulla nebulosa* Gld. coils up her yellow strings on the grassy flats, and *Haminea virescens* Sby. chooses the same place and time, but has a different shade of yellow for her egg-strings.

I was much interested in the eggs of *Actæon* (*Rictaxis*) *punctocœlatus* Cpr.

This mollusk has been rare, and I am inclined to think it only comes inshore in numbers during the breeding season and after that burrows in sand in deeper water for the rest of the year. In July we found them by the hundred.

The eggs are laid in a white string three or four inches long that coils so as to form a loose spiral.

The spirals are anchored, by some means, so firmly that the washing of rough surf does not sweep them away.

They so closely resemble the spiral pattern on the adult shell that the collector, looking down through the water, not unfrequently stoops to pick up what he thinks is one of these little gasteropods and finds a string of eggs in his fingers.

I visited Portuguese Bend and learned that *Purpura emarginata* Desh., which I found in quantity more than a year ago, is a resident or a comer and a goer, for more than a dozen were collected this summer. Its habitat is limited to a small mussel bed.

Other localities so much like this mussel bed, that one would consider them suitable dwelling places do not boast of a single Purpura; so that something besides collectors must disturb this usually common species.

I collected at San Pedro an abundance of *Acmœa paleacea* Gld. on the eel grass.

These close clingers love the grass on the outside of the island that is swept by heavy swells and where the water scarcely leaves them even in very low tides.

Their more peaceful cousins *Acmœa depicta* Gld. will probably be found swaying with the grass in the stiller waters of the bay, for dead shells have been frequently found in the drift.

In the quiet bay quantities of drift material are washed up with algæ and eel grass during medium tides.

This is rich in minute forms. It consists largely of broken shells of molluscs and crustaceans, but there is a sufficient quantity of *Pedipes, Siphodentalium, Tornatina, Cæcum, Truncatella, Mitromorpha, Turbonilla, Cerithiopsis, Triforis, Diala, Mumiola* and other wee bodies to amply repay any one for carrying away a few pounds of the drift to be dried and sorted at home.

The sifting and the sorting with a microscope takes so much time and patience, that the new and rare species hidden in my bags of drift must wait a more convenient season.

The yearly extension of sand flats at San Pedro, must make happy all sand loving species such as *Bulla, Sigaretus, Natica, Olivella* and scores of bivalves.

Besides these sandy stretches there are mud flats, rocky points, brackish water, fresh water, smooth or rocky beaches enough to make San Pedro an ideal collecting ground.

Although nearly all the localities are easy of access for the Conchologist, or the collector who "makes shell flowers," there are changes enough taking place to insure a good supply of shells.

COLLECTING ON AN ABALONE

BY F. W. KELSEY

Some of my young friends who collect shells at the seashore may be interested in the following method of getting specimens for a collection, when better means are not at hand.

On May fourteenth of this year I took a boat trip to the Coronado Islands, in Mexican waters, about twenty miles southwest of San Diego. Arriving at the anchorage at high tide, shore collecting was out of the question, so I went out with the skipper and mate in a glass-bottomed boat to a portion of the cove known as the "Marine Gardens". The water is very clear and at points where it is from two to three fathoms deep the view of the waving kelp, sea moss, grasses, shells and many colored fish is exceedingly interesting.

With a long-handled trident, or spear, the skipper would occasionally dislodge an abalone from the rocks, turn it over on its back and with a prong of the spear pierce the flesh of the mollusk and bring it up to the boat. About a dozen fine specimens were thus obtained, one being *Haliotis corrugata* Gray and all the remainder *Haliotis fulgens* Phil. The backs of several shells were covered with moss and other growths which I removed with my pocket knife from the backs of seven shells to be brought home for examination. The scrapings were treated to an all-night bath in a three-percent solution of formaldehyde, then rinsed and thoroughly dried, when they were shaken out and carefully examined for shells. From the material scraped from the seven shells I picked ninety-four specimens, including the twenty-five species which follow.

Amphissa versicolor Dall.
Assiminea californica Cooper.
Cerithiopsis columna Cpr.
Lasea rubra Mont.
Lacuna unifasciata Cpr.

Crepidula dorsata Brod.
Columbella aurantiaca Dall.
Columbella gausapata Gld.
Eulithidium substriatum Cpr.
Acmaea paleacea Gld.

Lacuna solidula Loven.
Mangilia striosa C. B. Ads.
Littorina planaxis Nutt. (juv.).
Odostomia americana D. & B.
Odostomia tenuisculpta Cpr.
Phasianella compta pulloides Cpr.
Fissurella volcano crucifera Dall.
Pecten, sp. (juv.).

Acmaea rosacea Cpr.
Acmaea asmi Midd.
Saxicava rugosa Linn.
Philobrya setosa Cpr.
Marginella regularis Cpr.
Psephis tantilla Gld. (1 valve).
Cardita subquadrata Cpr.

CRUISING AND COLLECTING OFF THE COAST OF LOWER CALIFORNIA.

BY FRED. BAKER, M. D., SAN DIEGO, CAL.

Fred Baker and his wife, Charlotte, were both practicing physicians in San Diego, California, during the early 1900's. Fred, born Jan. 29, 1854, in Norwalk, Ohio, was a graduate of the University of Michigan. He was active in local California civic affairs and was President of the San Diego City Council and a member of the Board of Education. In 1911, he served as malacologist and surgeon on the Stanford University expedition to Brazil. He contributed 15 articles to The Nautilus. *His large collection of shells was given to the San Diego Museum of Natural History soon after his death on May 16, 1938, at the age of 83. He was one of the most informed amateur conchologists of the Pacific Coast prior to World War I.*

Cruising on our southern Pacific coast is less indulged in than along the Atlantic seaboard, because there is a marked dearth of the land-locked harbors into which our eastern yachtsmen can run almost every night, or in case of a threatened storm. Nevertheless, two years ago, tempted by our summer promise of continued good weather, a party of seven, including my wife and two children, started from San Diego harbor for a run down the coast of Lower California in the staunch little schooner "Lura."

A late start made it advisable to anchor over night at the mouth of the harbor, but this gave a chance to get under way at daylight for a beautiful run of seventy miles to " Todos Santos " bay, on the sloping shores of which lies " Ensenada," the capital of the northern department of the Mexican territory of " Baja California."

As we ran we left broad to the starboard the Coronados, a group of seven small islands belonging to Mexico, but lying only twenty miles off San Diego, and a common terminus of our short cruises. They, like most of the off-shore islands, are bold volcanic masses, the largest, though less than three miles long, rising 880 feet above the sea, in many places sheer for hundreds of feet. This is a type of all the coast line for several hundred miles south. Bluffs and headlands, with here and there a narrow or broad valley sweeping down to the sea, but above all and crowning all, the foot-hills and the great mountains of the Coast Range.

It was just turning dusk as we rounded Ensenada Point into Todos Santos Bay, which is little better than an open roadstead, except for the protection offered by the chain of Todos Santos islands a dozen miles to sea, and the shelter of the Point from northwest winds. Immediately on dropping anchor we were boarded by the Comandante of the port, Don Luis Fernandez, and the quarantine officer, our old friend, Dr. Peterson, who courteously waived all examination, allowing us to go ashore at will. The two nights spent here with a nearly full moon shining down on us, just enough ground swell to keep in mind that we were cruising, and the balmy breeze of semi-tropical summer blowing over us, make a memory picture as near perfection as this world gives.

The day was busy. First we had the usual difficulties with the Mexican officials. In the absence of specific instructions, they were unable to determine whether we should register our craft as a private yacht or a passenger vessel. In either case they notified us that we must bear the expense and delay of telegraphing to the City of Mexico for instructions and license. Fortunately our schooner had on a former occasion been used in fishing down the coast, and after much argument Señor Victorio decided to grant us a three months' fishing license, at the same time clearing us with a clean bill of health for the return trip to San Diego. Under this very satisfactory arrangement we could run down the coast as far as the jurisdiction of the northern department reaches—something like 300 miles—land where we chose, collect what we liked, and when we were ready, sail away home without touching again at "Ensenada."

While our sailing master was arranging all this, the rest of us passed the day in seeing the few sights of the town, observing Mexican life, and visiting a few old friends. Among these Mrs. Gastelum holds first place, not for her society alone, though she is a woman of wide experience and much knowledge, but because in a former phase of her existence she was married to a Mexican customs official, who, during his sojourn at various ports on the Pacific Coast, had collected many bushels of shells which she has stored away in many boxes and barrels. This was the second time I had overhauled the lot, and as before I was astonished at the low price placed on my pickings, after a long conference between herself and her husband—a later acquisition. As I paid the bill I reflected that while I should undoubtedly have enjoyed the society of the former husband with his evident love of shells, the later acquisition was probably more in harmony with the size of my pocket-book.

Away at daybreak Sunday morning, looking our last on "Ensenada," one of the goodliest sights to look upon it has been my fortune to see in a fair amount of knocking about. A great sweep of unbroken sand beach from "Ensenada Point" to "Punta Banda," a distance of eighteen miles, the high range of Punta Banda breaking off abruptly into the sea to the south, the horseshoe being completed by the low mesa-crowned Todos Santos islands. The town of Ensenada nestles on the low beach under the high ridge which forms Ensenada Point to the northwest, the broad valley reaching back with few breaks for twenty miles—then the foothills, and back of all, as always, the great mountains! It is our dry season and everything is parched and brown, and the near-by ridges show great outcroppings of black volcanic rock, but the blending of color under our brilliant California sun, and the foreshortening of great distances giving the effect of haze and softness, make a scene of marvelous beauty.

A glorious sail—free with the prevalent northwest wind—out through the narrow gate between Punta Banda and the easterly island of the Todos Santos group, which was alive with seal and waterfowl, and down a bold coast for twelve miles to cast anchor under the lee of the "Santo Tomas" headland noted all along the coast for its frequent storms. Here we divided up, one to sleep, two to fish, two to hunt deer, and two to collect shells and algae. All were successful but the deer-hunters. Unfortunately I did not keep my Santo Tomas collections apart from others, so I can give no fair idea of my catch, but a single *Haliotis rufescens*, Swainson, represents the only species not appearing in the list which closes this article.

Away again at sunrise for our final southward stretch. All day we ran almost before the wind, the coast growing generally more bold and culminating in Cape " Colnet," a great promontory presenting an almost unbroken face to the northwest, a cliff many miles long and many hundred feet high. We round the Cape with a half gale, and bear away southeasterly to our final destination, the little island of San Martin, lying five miles off the coast and ten miles from San Quintin, the first land-locked harbor in 200 miles from San Diego. We cast anchor at 3 a. m., and all hands slept late.

Of San Martin a few words' description must suffice. Roughly it is a round conical island, three miles in diameter, with two peaks, the higher a typical extinct volcano rising 471 ft., with an almost perfectly regular crater about 250 ft. in diameter, and between 75 and 100 ft. deep. The island is a solid mass of very hard volcanic rock with frequent small caves—evidently blow-holes—covered imperfectly where reasonably level by a thin soil which supports a moderately abundant vegetation in which various species of cactus are very plentiful. Up the slopes are great slides of loose rock, and owing to the cacti and the roughness of the way, the climb of a little over a mile to the top proved a very serious undertaking.

On the north side of the island a moderately level space, covering between 500 and 1,000 acres, is occupied by rookeries, mostly of pelicans and cormorants. The birds were most of them just beginning to fly, and a rough estimate convinced us that there were certainly some millions of them. We spent the greater part of one day watching them. The young cormorants waddled to the bluffs, spread their wings evidently for their first trial, and sailed or flew awkwardly into the ocean. There they were perfectly at home and could not be distinguished from the old birds, swimming and diving with perfect ease. But the pelicans had a harder time. They could fly very well indeed, but like the Irishman " had a divil of a toime loighting." Starting from some slight elevation they would sail away majestically, managing their great wings and bodies remarkably well. After a turn of one or two hundred yards they would light without slowing up perceptibly, come down with a thud that we could hear a hundred yards away ; turn two or three somersaults, and straighten up with the same appearance of surprise and offended dignity which we have all seen drunken men assume when suffering from similar mishaps. We actually laughed till we cried, and it was hours past our dinner time before we could agree unanimously to start for the boat.

Running easterly at a tangent from the southerly edge of the island for nearly 1000 yards is the so-called breakwater, a nearly straight line of enormous beach-worn boulders arranged like some huge artificial jetty. The acute angle has filled in with sand over a space of about fifty acres. In the bight there is safe anchorage except in a northeast storm. At two places dips in the breakwater bring it below high tide level, one opposite the little harbor, and the other opposite the sand bar, and here the constant tidal current has excavated a little circular bay, covering two or three acres. This bay and the breakwater, with another little bight not much over thirty feet across, furnished nearly the only good collecting ground on the island. Otherwise I found only a few of those hardy shells capable of standing any amount of buffeting by the waves, Chitons and Limpets, an occasional Chlorostoma, and the *Monoceros lugubre* Sby., which is in evidence along the whole coast from Ensenada south.

Dredging at moderate depths gave little results. but some of my fishermen friends who spend much time about this island, which furnishes some of the great fishing of the coast, make a practice of bringing up to me rocks which they haul up on their lines from considerable depths, attached to kelp roots. I am, therefore, able to list a considerable number of deep-water species. It has seemed to me advisable to publish the following list of shells secured from this small island and its immediate vicinity as a contribution to our knowledge of geographical distribution. I have to thank Dr. Wm. H. Dall of the National Museum, and Mr. Henry Hemphill of San Diego, Cal., for determining a very large share of the species about which I was in doubt.

After commenting on the fact that many of the specimens which I sent to him were too young or too worn to be identified specifically, Dr. Dall writes, " There was a small *Rissoina* among the shells which we have had for some years from San Pedro, but had not named, and with your permission we propose to call it R. Bakeri, Dall and Bartsch. There are also some of the new *Pyramidellidæ* described in the paper on W. Am. *Pyramidellidæ* which Mr. Bartsch and I have in preparation."

Our return trip was made much more slowly than the outward one, as the prevalent wind made it a long tack to windward. The only break was a night run against a sharp storm to make the doubtful shelter of Santo Tomas, where we lay for twenty-four hours with two anchors out, estimating the chances of a shift of the wind driving us to sea again. Our cruise lasted seventeen days, and was unanimously voted a success.

SHELL COLLECTING IN PUGET SOUND AND ALASKA.

BY DR. FRED BAKER, SAN DIEGO, CAL.

During the last summer it was my good fortune to take a most enjoyable trip to Seattle, Washington, where the Alaska-Yukon-Pacific Fair was in full swing; to the San Juan Islands in Puget Sound, where the Summer School of the University of Washington was in session; and on to Alaska, reaching a point as far west as Cook's Inlet. At all points possible I collected shells, and a full list of species and localities follows at the end of this article. It is a grateful task to acknowledge help in naming all doubtful species at the hands of Dr. W. H. Dall and Dr. Paul Bartsch of the National Museum, and the Rev. Geo. W. Taylor of Nanaimo, British Columbia. Without their help this article could only have been an account of the wanderings of a very amateurish conchologist.

The University of Washington, in connection with several other institutions which join forces with her, conducts a summer school on a group of islands in Puget Sound,—there are a hundred and fifty of them which constitute the county of San Juan in the State of Washington,—and here my wife and I dropped down on about fifty of the best people on earth, as biologists always are. This year they were trying the experiment of doing work in two places, and we arrived just in time to help them move camp from San Juan Island to Olga, on Orcas Island, thirteen miles away.

Orcas Island, the second largest of the group, supports a considerable farming population, but is much broken, and boasts a peak, Mt. Constitution, 2660 feet high, from which we had a glorious panoramic view in all directions over thousands of square miles of water with scores of islands, and the mainland rising foothill above foothill to the Olympics and the Selkirks with their snowy ranges, and dominating and crowning all, the splendid peaks of Mt. Baker and Mt. Rainier.

We were soon established in camp quarters after our arrival, and regular work got under way. Here I met for the first time Dr Kellogg of Williams College, who was doing some most interesting work on the siphonal currents of the pelecypods. Others were doing equally interesting original work in various lines of botany and zoology, but of course the dredging was most important from my point of view. Their dredging was done by a regular shrimp dredger who was employed by the season. The apparatus is crude but very efficient. A trawl twelve feet wide was used, and as the water is shallow everywhere, the boat only carried about fifty fathoms of cable. No soundings were taken, but the depths, which generally varied between ten and thirty fathoms, were ascertained from the charts. I have done considerable work with the Alexander Agassiz, the boat of the Marine Biological Association of San Diego, and I spent five days aboard the Fish Commission Steamer Albatross off San Diego, but I have never seen such a wealth of material as we got at all stations. The great trawl always filled in a few minutes, but I was impressed by the fact that the variety of species was not nearly so great as in our more southern waters. As no one was specially interested in mollusks, I was allowed the privilege of taking any of the dredgings which I could handle. In fact we were royally treated by everybody at the station, and I shall always count the ten days we spent there as " the best yet."

A week later we were on our way to Alaska, that country of which some one has said, " If you are old, go by all means; but if you are young, wait. The scenery of Alaska is much grander than anything else of the kind in the world, and it is not well to dull one's capacity by seeing the finest first." As representatives of one of our local papers we were a part of the National Editorial Association, and as far as we went all Alaska was ours for the asking. The usual Alaska trip consists of a run from Seattle up to Skagway and back, the whole distance with the exception of a dozen miles or so being in land-locked channels, often only a few hundred yards wide, bounded by steep, heavily wooded mountains on either hand. Various towns of much interest are visited, and a side trip up Taku Inlet shows the great Taku and Windom Glaciers. Formerly Glacier Bay was visited for a view of the Muir Glacier, but the immense amount of float ice and icebergs which have formed within the last two or three years make this visit unsafe and it has been abandoned.

On our trip we saw all that the regular tourist sees, and for full measure we ran west from Juneau into the open ocean, making a run of several hundred miles to Prince William Sound and Cook's Inlet. As we ran out of the still waters of Cross Sound into the open ocean early one morning on our right opened up the Fairweather range, one of the finest mountain ranges in all Alaska. One after another the great peaks came into view, until the whole range was off our starboard side. Many mountain ranges do not show their full extent, as to be seen the foothills must be crossed, and one's own elevation cuts down the apparent height. Here the range comes to the very ocean, and the overpowering sense of altitude is there. Four peaks rise above 10,000 feet, and Mt. Fairweather towers to 15,292 feet. From one point we could count six great glaciers, making their beginnings in the everlasting snows at the top and winding down like great rivers toward the sea.

At Cordova, on Prince William Sound, a place of 1200 people, which has grown in a year, we were taken inland fifty-three miles on the Copper River Railroad. This road, which is ultimately to reach a veritable mountain of copper, crosses several branches of the Copper River which are making the delta at its mouth, and then goes up the left bank until a point is reached where on each side, and within four miles of each other, a great glacier comes down to discharge its icebergs into the river. The face of each glacier rises sheer 350 to 400 feet, and each is over four miles wide along the river's bank. Opposite the lower, Child's Glacier, at a distance of 1500 or 2000 feet, we amused ourselves taking photographs, and repeatedly were successful in catching the great masses of ice as they broke away. In the face of this body of ice we actually suffered from the heat of the June day.

At Cordova I took advantage of a fine tide to do a couple of hours collecting, being rewarded by several good things, the best being a fine *Mactra (Spisula) alaskana* Dall. From Cordova we went to Valdez, the point of departure of the great dog teams for the Fairbanks region during the winter, but a very dull town in summer.

A year ago $700,000 worth of gold was brought out on dog sledges. Unfortunately I was not able to do much collecting here, as it is almost imperative to take a small boat and get away from the waterfront of the town, and there was not sufficient time.

On the return trip we called at Ellamar, a very small mining town on a very small landlocked bay off Prince William Sound, where we had to run in on one high tide and wait for the next to take the ship out. I therefore had my work cut out for me, and, on account of the steady downpour of rain, I donned waterproof and rubber boots, and protected by an umbrella from the heaviest rain, I made the most varied and interesting catch of the Alaskan portion of my trip.

From Prince William Sound we sailed at midnight to Seward on Resurrection Bay. When I turned out in the early morning I found the tide well out and still going down, so I got busy at once. Here I found many things of interest, the best being *Trachydermon raymondi* Pilsbry, of which I succeeded in finding about a dozen. I finished my work well before the call to breakfast, and then we had a fine ride of twenty-five miles over the Alaska Central Railroad, which is now in the hands of a receiver, but was then being investigated by Mr. Geo. W. Perkins, of the firm of J. P. Morgan & Co., with a view to syndicating its completion. Since our return the papers have announced that this would be undertaken in the near future. The road already runs fifty miles into the interior through a region of marvelous beauty, and it has one unique feature of construction. At one point the engineers have blasted out a deep pass directly through a great glacier, which in future will have to be constantly cut away, as it encroaches on the railroad right of way.

From Seward we sailed down Resurrection Bay on a glorious afternoon, passing close to the rugged group of islands at the mouth, and looking back up the whole length of a great white glacier directly to the setting sun which was painting everything in matchless reds. Then the open ocean, and heavy clouds, and night settling rapidly, and cold winds driving us to our bunks for a much-needed rest.

Morning found us in Cook's Inlet, our most westerly point, and we sailed into the little station of Port Graham just at daylight. As we ran into the small bay where this station lies, we saw across the broad inlet the smoke of St. Augustine, the only volcano we saw which was even active enough to smoke. Aside from this slight interest, it is a most beautiful mountain, a nearly regular cone rising almost from the water's edge for many thousand feet, and covered with glaring white snow from very near the base.

We spent the whole day at Port Graham, and the tide had run a long way out when we tied up. So it was overalls and rubber boots and a fine hunt of a couple of hours before breakfast. Then a hurried breakfast and three hours more of hunting. The collecting was good everywhere, so I could begin beneath the wharf and work either way. I found *Argobuccinum (Priene) oregonensis* Redfield, in quantity on the beach, and their numbers made me miss an important possibility. I had secured a considerable number of these shells in dredgings in Puget Sound, but the temptation to take more, which I suppose is the miserly instinct common to most collectors, induced me to throw a lot of them above the tide-line to be picked up later. When the tide had come up beyond the limits of profitable collecting I began to gather in this bunch of shells. Then I found among them four fine specimens of *Chrysodomus liratus* Martyn, which I had overlooked in my hurry, though they are so different that I cannot understand my blunder. It would have been more profitable to have hunted the beaches over for this last species, of which I have little doubt that I could have found as many as I could have cleaned for transportation.

At Port Graham I also found in quantity *Astarte rollandi* Bernardi. This fine shell I had not seen before, and I was glad to add it to my collection. The Indians were running around on the beach and in their bydarkas, and some of them watched me cleaning my catch. I have no doubt that my actions were a puzzle to them. None of them could be induced to talk any English, even if they knew any. However, by laying down ten cents and then tying an *Astarte*, then making use of a sign language invented for the occasion, I was able to interest a small native, and I finally kept him tying my bivalves for a couple of hours. With his help my day's work ran from 4 a. m. to 4 p. m., with very short intermissions for breakfast and lunch, so that on this day at least I felt very tired and very virtuous and slept the sleep of the just during the long night following.

Every hour of our 8000 miles of water travel was most enjoyable, as were the intervening stops, and from the point of view of a collector the trip was eminently profitable.

SHELLS OF LA JOLLA, CALIFORNIA.

BY C. R. ORCUTT.

Charles Russell Orcutt, an amateur botanist, conchologist, and editor of a nature magazine, was born in 1864 in Vermont, and went West in 1879. He was an ardent collector and attempted to get various museums to subsidize his field trips with varying degrees of success. He collected in California for 30 years, then in Mexico, Central America, Jamaica and finally Haiti where he died on August 24, 1929, at the age of 65. His collections went to the U.S. National Museum in Washington, D.C.

My acquaintance with La Jolla dates back to 1879, when there was not a human habitation on the coast from San Dieguito on the north to the old lighthouse, 500 feet above the beach, at the extremity of Point Loma at the entrance to San Diego Bay. Now there are several flourishing towns along the way, the delight of summer and winter tourists, among whom not a few have been conchologists.

Taking charge of Hotel Strand at La Jolla in July, 1918, I have since busied myself quite as much with the molluscan fauna of La Jolla as with the hotel business, with some interesting results.

Mr. Maxwell Smith has contributed a list of La Jolla shells to THE NAUTILUS (volume 21, pages 55 and 65), and Mr. Joshua L. Bailey, Jr., has contributed a supplementary list (on page 92). A few additional notes may be of interest.

Haliotis fulgens.—In the spring of 1916 San Diego was visited with great floods, and a great amount of silt was washed into the ocean via San Diego and False Bay with the fresh water. This proved to be fatal to many mollusks, and I am told that many thousands of this shell were washed up along the shore from False Bay to La Jolla. One gentleman told me that a train of cars could have been filled with these shells which were piled a foot deep on the beach in many places. Another filled two sacks with the shells and nearly broke his back tugging them to the top of the cliff at what in early days we called Seal Rock, now named Bird Rock Beach. These he has finally placed at my disposal, and I found the two sacks full chiefly of this species. *Haliotis cracherodii* and *H. rufescens* were missing, as well as *H. assimilis*.

A SUMMER'S COLLECTION AT FRIDAY HARBOR, WASHINGTON.

BY T. S. OLDROYD.

Tom Shaw Oldroyd and his wife Ida Shepard Oldroyd both came to California in the 1880's. Tom was born in England in 1853, Ida in Goshen, Indiana, in 1856. They became acquainted through their common interests in shells, and were married in 1895. Their extensive collection was sold to Stanford University in 1917 and with it came Mrs. Oldroyd's curatorship until her death in 1940. She published several large tomes, including the useful and well-known compilation, "The Marine Shells of the West Coast of North America." Tom contributed six articles to The Nautilus. *He died on Nov. 3, 1932, at the age of 79, and Ida, an Honorary President of the American Malacological Union from 1935 to 1940, died July 9, 1940, at the age of 84.*

Mrs. Oldroyd and myself had the pleasure of spending our vacation at the marine biological station of the University of Washington at Friday Harbor, San Juan Co., Washington. This group of islands, so wonderful in their wild beauty, is situated between the Strait of Fuca and the Strait of Georgia, north of Port Townsend. The San Juan group comprises more than 100 islands of varying sizes, the most important being San Juan,

noted as having been the scene of the last struggle between the
British and Americans in the boundary-line dispute from 1852
to 1872; the ruins of the old English camp and blockhouse still
remain near Roach Harbor. Friday Harbor, the chief town of
the islands, is the county seat of San Juan Co. which includes
all the islands of the group. The town is situated on a beautiful bay and is the shipping-point for a large area of unsurpassed
agricultural land. They have also one or two large salmon
canneries. The islands are nearly all high and prominent and
covered with a dense growth of trees, mostly fir. Mount Constitution on Orcus Island is the highest point. From its summit, 2094 feet above the sea, is the finest view to be obtained
anywhere of the great panoramic picture of Puget Sound. The
biological station is about one-quarter mile from Friday Harbor,
and is under the able management of Dr. T. C. Frye. The
camp is situated on a beautiful picturesque heavy-timbered
slope and is one of the most healthy places in the world. The
islands are a paradise for the botanist and student, as all forms
of marine life is here very abundant. Students and visitors are
accommodated from all over; tent houses, cots, mattresses, and
other things are furnished, all but the bedding; and one needs
plenty of covering for the nights are cold. Good table board
can be had at nominal rates and visitors are required to pay a
small registration fee which entitles them to the use of the boats
and the week-end excursions on the dredge boat to places of
interest on the islands, and it is well worth the price. The excursions sometimes take two days with a camp out over night,
and they have bonfire chats and clambakes in the evenings.
Sometimes in the main channels the tide runs swift, but in the
bays and protected places and along shore it is as smooth as a
mill pond. Although there is a difference of 14 feet sometimes
in the tides yet it creeps in and out without a splash or a ripple.
This makes it fine for shore collecting on the rocky reefs. It
is the best I ever saw. The dredging was done by a small tugboat, a shrimp dredger, and Captain Burnham understood the
business thoroughly, having been 25 years on the Sound. He
always knew the character of the bottom in nearly every place.
We were allowed to go out on the dredge boat every day. At
this we were treated especially fine, there being nobody there

very much interested in shells, and often for two or three days at a time when they had no special use for the dredge they turned the boat over to us to go dredging where we pleased, an opportunity we were not slow to grasp and make good use of. And just imagine our having to sweep overboard bushels and bushels of those beautiful *Chlamys hericius* Gld. and *hindsii* Cpr. dredged in from 25 to 50 fathoms, not knowing what to do with so many. Mrs. Oldroyd worked like a beaver all the time and did not let many good things get away. The summer school lasted six weeks and a pleasanter time we never spent.

NOTES OF A COLLECTING TRIP TO DEPARTURE BAY, VANCOUVER ISLAND.

BY GEORGE W. TAYLOR.

The richness of the northwest Pacific area of North America is evident in Reverend George W. Taylor's account of his collecting experiences in 1894 on Vancouver Island, British Columbia. It was at the Bay of Departure, near Nanaimo, that Taylor was asked to establish and run a biological station by the Canadian Government. He published several good lists of British Columbia marine mollusks. His collection was acquired by the Oldroyds who, in turn, sold it to Stanford University. His freshwater mussel collection is at present at the Delaware Museum of Natural History. Taylor contributed eleven articles to The Nautilus. *Unfortunately, historians have not recorded the details of his life, other than that he was born in 1851 and died at his home on Departure Bay in August 1912.*

The account given by Mr. Wood in the October number of THE NAUTILUS of his collecting trip to Monterey Bay tempts me to put on record an experience of my own which goes to prove that our Northern waters are quite as rich in molluscan life as those of the sunnier South.

Departure Bay is a small bay on the east coast of Vancouver Island, about 75 miles north of Victoria. It is shut in by its own shores on the north, west, and south, and is protected from the open sea on the east by a series of small islands. Consequently, the water, at most seasons of the year, is smooth, and as the depth ranges from 10 to 50 fathoms, and the bottom is varied, being sandy in some places and rocky in others, the bay is a capital place for a dredging expedition.

In August, 1888, through the kindness of Mr. S. M. Robins, the managing director of the New Vancouver Coal Company, I was able to spend four days in dredging from a small steam launch belonging to the Colliery Company. Much time was wasted on this occasion in searching for suitable ground, but the result of the four days' work was by no means disappointing, as I took home with me more than five thousand (5,000) specimens of one hundred and ten different species. One of them, since named in manuscript by Mr. Whiteaves as *Pecten Vancouverensis*, was new to science, and several others were additions to our Vancouver lists.

In July of the present year, I spent three and one-half days in the same locality in company with Professor Macown, the well-known Canadian botanist and naturalist. We were determined, if possible, to beat the previous record, and therefore worked very industriously. We spent two and one-half days collecting on shore between tide marks, and one day was devoted to dredging over the ground prospected in 1888.

In the shore collecting my own captures amounted to nearly 2,500 shells of 61 species. In the dredging expedition our joint bag reached the grand total of over seven thousand (7,000) specimens of 88 species. This very satisfactory result was obtained from an ordinary open sailing boat, with a crew of three men and a single home-made dredge. I should mention that 13 species were taken both between tides and by dredging, so that the actual number of different species taken in the three and one-half days was 136.

In shore collecting, no attempt was made to collect the very common shells in quantity, or the numbers might have been swelled indefinitely. In fact, I was looking more especially for Chitons, of which I took seven species (183 specimens), and the smaller shells, such as *Volutella pyriformis* Cpr. (40 specimens), and species of *Odostomia*, which occurs commonly under rocks at low water. I took also on this occasion a very fine series of *Terebratella trans-*

versa Sby., which was found literally in thousands attached to the rocks on the south side of the bay.

Fine series of several *Macomas* were dug in the sand, and a few specimens of the curious *Lepton rude* Whiteaves were found attached (as is their habit) to the central segments of specimens of *Gebia pugetensis*, which we dug out of the muddy shores.

When dredging, everything that came up was preserved, and the finer sand and mud boxed, and afterward dried and examined at home.

Of the Pelecypoda dredged, by far the commonest specimen was *Acila Lyalli*, of which at least a couple of thousand specimens were taken. Another common bivalve was *Cryptodon sericatus* Cpr., about 300 specimens. *Venericardia borealis* Conrad, *Nucula tenuis* Mont., and *Lucina tenuisculpta* Cpr., came next in order, about 100 of each being taken.

Of rarer shells may be mentioned, *Pecten Vancouverensis* Whiteaves and *hastatus* Sby., *Amusium caurinum* Gould, *Modiolaria lævigata* Gray and *nigra* Gray, *Crenella decussata* Mont., *Cardium blandum* Gould, *Fulvia modesta* Ad. and Rue., *Cryptodon flexuosus* Mont., *Astarte Esquimalti* Baird, and *undata* Gould, *Psephis Lordi* Baird, *Clementia subdiaphana* Cpr., *Venus Kennerleyi* Reeve, *Macoma yoldiformis* Cpr., *Cuspidaria pectinata* Cpr., *Kennerlia filosa* Cpr., etc.

Of the Gasteropoda the most abundant were, *Nassa mendica* Gould, *Nitidella Gouldii* Cpr., *Olivella baetica* Cpr., *Mesalia reticulata* Mighels, and the deep-water variety of *Margarita pupilla* Gould, of each of which more than one hundred specimens were obtained.

Of rarer shells the following is a partial list: *Drillia incisa* Cpr., and *cancellata* Cpr., *Mangilia sculpturata* Dall, *Cancellaria circumcincta* Dall, *Velutina laevigata* Linn., *Turbonilla torquata* Gould, *chocolata* Cpr., and *Lordi* Smith, *Scala indianorum* Cpr., *Solariella peramabilis* Cpr., and *varicosa* Mighels, *Puncturella galeata* Gould, *cucullata* Gould, and *Cooperi* Cpr. (all living), *Cryptobranchia concentrica* Midd., *Ischnochiton interstinctus* Gould, and *cancellatus* Sby., *Placiphorella sinuata* Cpr., *Utriculus incultus* Gould, and *Rictaxis punctocælata* Cpr., this last being new to our Vancouver list.

I have not attempted to give a complete list, as such would take up too much of THE NAUTILUS's valuable space, and would, moreover, be of little interest; but I think I have written enough to show how very abundant the Mollusca are in our seas, and how much may be accomplished in even a single day's collecting if one knows exactly how and where to look.

MARINE SHELLS OF DRIER BAY, KNIGHT ISLAND, PRINCE WILLIAM SOUND, ALASKA.

BY WALTER EYERDAM

Walter Jakob Eyerdam, born in Seattle in 1892, was an incurable naturalist who collected omnivorously around the world. He has been acclaimed as a superior collector by botanists, ornithologists and conchologists. For years he made his living making wooden barrels for Alaskan whaling and herring stations. His exotic travels took him to Siberia five times, to the Solomon Islands on the Whitney-American Museum expedition of 1929, to Alaska for 25 summers, and to Chile and Haiti. He was well-known and admired among the leading scientific circles of America for half a century. Many new species were named in his honor. Eyerdam died on Dec. 31, 1974, after a long illness, in Seattle at the age of 81. His collecting energies are evident in the following article which was one of 35 that he contributed between 1924 and 1961.

During the summer of 1923, while engaged at my trade as cooper in the herring saltery of the Knight Island Packing Co. in Drier Bay, Prince William Sound, I utilized much of my spare time in dredging and searching for shells along the beaches.

Drier Bay is about 35 miles by boat from La-Touche; it is an indentation of Knight Island, about 10 square miles in area. Steep mountains surround the bay, causing steep rocky beaches in most places and often fairly deep water in close to the shore. In Mallard Bay, Cat-head Bay and Miner's Cove, as well as several other coves and small bays belonging to Drier Bay proper, the beaches are covered with fucus and at low tide with abundance of eelgrass, and the water is quite shallow.

The best collecting was obtained by washing the roots and leaves of eelgress, which often grew to the length of 12 feet. About a ton of eelgrass was washed in tubs and the mud screened and panned, after which the remaining débris and shells were carefully examined and separated, and many rare species of small shells were thus obtained.

Most of the Odostomias and several other small rare species were found among stones and sea-moss at low tide in Mallard Bay. About a dozen species of Chitons were found, some of them from a depth of 200 feet.

Dredging was rather poor considering the large amount of physical effort that was performed. About 150 hauls were made at depths of 10 to 40 fathoms. Most of the dredging was done near the herring dock by rowing out about 200 feet and dropping the dredge and then rowing back to tie the boat to a piling, after which I proceeded to haul in the dredge rope by hand. It was generally chockfull of rocks and débris and often weighed over 200 lbs. The results were not always gratifying considering the meager returns that I generally got, but whenever I found a new shell I would try again. Sometimes in the fall of the year when there was very little work to do in the shop I would spend most of a day out in a small rowboat dredging shells. Three times during the season I lost my whole dredging outfit in waters ranging from 35 to 200 feet in depth, but each time I succeeded, with patience and perseverance, to snag it up again.

Many rare small shells were dredged, including two new species: *Turbonilla (Pyrgolampros) eyerdami* (Bartsch) and *Puncturella eyerdami* (Dall), and twenty-six species were found beyond their geographical ranges.

SUBSCRIBERS.

A blank bill enclosed means that your subscription has expired and should be renewed at once. Please remit either by money order or draft on New York City or Philadelphia banks. Do not send personal checks outside of New England and the above cities.—EDITORS.

COLLECTING SHELLS IN THE KLONDIKE COUNTRY.

BY P. B. RANDOLPH.

Little is known today of this amateur collector, P. B. Randolph, who lived in Seattle, Washington, during the end of the last century. He evidently was an ardent shell collector and corresponded with two of the leading mollusk men in the East, George Clapp and Henry Pilsbry. Randolph was obviously on an unsuccessful gold rush trip to the Klondike, and assuaged his financial disappointments by recounting his conchological triumphs. On this trip Randolph discovered three new species that were named after him by Pilsbry and by Vanatta. Much of the gains by goldseekers ended in the brothels and gambling halls of the seaboard towns, but at least Randolph's name is still ringing in the list of Alaskan species of mollusks.

We left Seattle, Wash., on August 1, 1897, for the Klondike gold fields. Our first stop was made at New Metaketta, Duncan's Island, Alaska. We only made a short stop here to take on water. I made a rush for shore, and, in a short time, had collected a few each of *Circinaria sportella hybrida* Anc., *Circinaria vancouverensis* Lea and *Polygyra columbiana* Lea. These were found under the logs and boards just above high tide mark. No further stop was made until we reached Dyea, at the head of Lynn canal. From here we had to be our own pack-horses to the lakes. The Dyea valley is heavily timbered and the narrow bottom land covered with alder. We laid over one day, about half-way to the Dyea cañon, and I improved the time collecting the small species found there, consisting of *Pyramidula striatella cronkhitei* Newc., very plentiful under dead leaves and sticks, *Conulus fulvus alaskensis* Pils., *Punctum conspectum* Bld., and *Pupa decora* Gould. This last was very plentiful, and I think that in one day's faithful collecting I could supply the cabinets of the world.

Packing 100 pounds over a pass 3,000 feet high did not tend to arouse my conchological ambition, but at each stop I prospected the dead leaves and sticks with varying success.

We laid over one day at Lake Linderman, resting from the past week's hard work, and I had time to hunt over the flat at the head of the lake, where a small stream empties in. Here I found several dead shells of the *Vitrina exilis* Morel., and was despairing of finding any alive, but at the last moment found three under a small dead stick. These were the first of this genus that I had ever seen alive, and I felt well paid for the time spent. I also found *Pyramidula striatella cronkhitei* Newc. and *Conulus fulvus alaskensis* Pils. Associated with these were two shells that Mr. George H. Clapp and Mr. H. A. Pilsbry considered new, and were kind enough to name in my honor, *Zonitoides randolphi* Pilsbry and Clapp. At the head of the lake, near our camp, the rocks were covered with *Valvata sincera* Say and *Limnæa palustris* Müll.

The next day we put together our canvas boat, made of twenty-ounce duck, ready for our 600 miles trip down the Lewes and Yukon rivers to Dawson. At Marsh lake found dead shells of *Limnæa ampla* Migh., some very large, one measuring one inch and a half long and one inch across, and a dwarf variety of *Limnæa palustris* Müll.

The only shell collected going down the Lewes river was *Succinea nuttalliana* Lea.

We ran the famous Miles cañon in our canvas boat, but packed our outfit and boat around the White Horse rapids.

I had no further opportunity for collecting until we reached Dawson, Northwest Territory. There I found dead shells of a *Succinea*, where a fire had run through the moss, but they were too fragile to handle.

Snow commenced to fall on September 12th, and that put an end to collecting trips.

We spent the winter on one of the claims on Bonanza creek, in the ordinary occupation of a miner in that latitude, which would be another story.

After the clean-up in the spring we rebuilt our canvas boat in the shape of a scow to go down the Yukon river 1,800 miles to St. Michael's. We left Dawson on June the 9th and leisurely floated with the current, enjoying the days twenty-four hours long; that is, at Fort Yukon the sun was visible all the time. As I heard one man ask another "the time of day," "Eight o'clock" was the answer. The first said: "I am worse off than before; I do not know whether it is night or morning."

I did not find any live shells on the upper river, but on the bars found a few dead shells of *Succinea chrysis* West.

The mosquitoes were very bad on the lower river, and it was nearly suicidal to go into the brush; but when about twenty miles below Andreafsky we were compelled to lay over on account of wind and rain. I tried the experiment of building a smudge in the gold-pan and carrying it with me. I was rewarded by finding that the ground and stalks of grass were alive with *Succinea chrysis* West., and before the day was done I had nearly a pint cup of them cleaned.

The next day we left the main river and followed a slough that led us into three large lakes that run to the foot of the mountains. The banks are ten to twenty feet high and perpendicular. Near the water was a stratum of shells (*Macoma inconspicua* Brod.), about four inches thick. This locality is about 100 miles from the Aphroon mouth of the Yukon.

At an Indian camp below Holy Cross Mission I saw the right valve of an Anodonta used as a spoon by an old squaw. She could not understand, or would not, so I could not learn where it was found. She also prized it so highly that, though offered a good trade, she would not part with it. It was the size and color of our *Anodonta oregonensis*.

We made an early camp at Point Romanoff, which is about half way from the mouth of the river to St. Michael's, on the Arctic Ocean. Here I had to make use of the same expedient that I used before to "stand off" the mosquitoes, and found on the drift wood on the beach specimens of *Conulus fulvus alaskensis* Pils. and *Pupa decora* Gould. After entering the canal that connects the Arctic Ocean with Norton's Sound I found the small ponds that are common on the trundra full of *Limnæa palustris* Müll., most of them of large size. I also found a small bivalve very plentiful that was new to me, and I collected a large number of them; but, alas! they belong to the Crustaceans and the other fellows are enjoying them.

Florida and West Indian Shells

For Sale and Exchange
Send lists of duplicates and desiderata.

N. W. LERMOND
P. O. Address: **Gulfport, Fla.**, until May 1
After that date, **Thomaston, Me.**

Tracking the Terrestrials

The pleasures of land shell collecting in the lush woods along the magnificent banks of the "Rhine of America" in Duchess County, New York, were recounted in several articles by William S. Teator, an amateur conchologist from Upper Red Hook, not far from Kingston. Teator corresponded and sent specimens for identification to W. G. Binney and Henry A. Pilsbry in the 1880's. The giant Succinea from Cruger's Island was described as a new subspecies (ovalis **optima**) by Pilsbry in 1908. The editor, H. A. Pilsbry, obtained permission from Dr. William D. Hartman of West Chester, Pa., to add the figures of land snails from the book, Conchologia Cestrica, published in 1874.

COLLECTING LAND SHELLS IN EASTERN NEW YORK.

BY W. S. TEATOR.

Following the names of the snails illustrated in Teator's article are given here the modern nomenclature from Pilsbry's Land Mollusca of North America (North of Mexico), *1939-1948.*

Helix albolabris *now* Triodopsis albolabris *(Say, 1816)*

Helix palliata *now* Triodopsis notata *(Deshayes, 1830)*

Helix thyroides *now* Mesodon thyroidus *(Say, 1816)*

Patula alternata *now* Anguispira alternata *(Say, 1816). This very common and widespread species was first illustrated in 1685 by Martin Lister, Physician to Queen Anne, his specimens having been sent from the Virginia Colony.*

Selenites concava *and* Macrocyclis concava *now* Haplotrema concavum *(Say, 1821).*

Zonites fulginosus *now* Mesomphix cupreus *(Rafinesque, 1831).*

Near the east shore of the Hudson, midway between Tivoli and Barrytown, in Duchess Co., New York, is Cruger's Island. It has an area of seventy-five acres, and is so richly endowed with beauties and attractions—nature's gifts, which the owners have carefully fostered—that to the visitor it seems a place of enchantment. The scenery is especially fine; an almost undisturbed view for miles up and down the "Rhine of America," with the majestic Catskills some ten miles distant to the west, a beautiful background to the picture; while from its many winding paths are ever-changing vistas of water, mountain and sky.

Tracking the Terrestrials

At the northeast a large stream, the White Clay Kill, rushes down the rocks through a romantic glen and has its outlet. South of this, for a long distance, fringing the east shore of the cave, and having a width varying from an eighth to a half mile, is an extent of heavily-wooded land of perhaps two hundred acres, part of a park-like domain of Revolutionary days called "Almont." The soil is of decided clayey character, and there are a half dozen little rivulets coming from the hills at the east running through to the river. With their numerous tributary branches they have cut their way down through the plastic earth making quite an intricate succession of deep gullies.

Here are hundreds of grand, massive white oaks, beeches, and hickories, growing so thickly as to almost shut out a glimpse of the sky. It is a scene of primitive sylvan grandeur not often found in this part of the country. Great numbers of fallen trees and decaying logs are lying in every ravine, and the ground is thickly carpeted with leaves. It is thus an ideal home for the land snails, which flourish in abundance, and a "happy hunting ground"

Helix albolabris.

for the enthusiastic collector, who, if he pays it a visit during a warm, humid day of summer—just after a shower for instance, when everything among the trees is saturated, and the air is smoking with moisture—will find the woods literally teeming with Molluscan life.

H. palliata.

The writer on one such day carried home actually two quarts of splendid live specimens in his pockets, besides having filled all his collecting boxes. They speedily became quite a slimy mass, not conducing in any great degree to personal comfort, but who among the Nautilus people could resist a like temptation?

At such a time an abundance of *Helix albolabris*, large and beautiful, and *H. thyroides*, crawling about the logs, and traveling among the leaves; plentiful supplies of *H. alternata* and *palliata*, but keeping nearer at home; ocasionally a *Zonites fuliginosus*,—

H. palliata.

a very pretty shell when perfect; many of *H. tridentata; H. monodon* (*fraterna*), and *hirsuta* to be had on closer search among the stones in the vicinity of the falls; while down at the river's edge, on the rushes and weeds, are thousands of *Succinea ovalis*, and associated with them though in greatly lessened proportion, is an elongated form of *S. avara* of dark amber color, some individuals of which are found reaching 11 millimeters in length.

Patula alternata.

More careful hunting under the logs will bring to light good specimens of *Zonites arboreus, indentatus,* and *viridulus;* the last two rather scarce. Also a few *Zonites fulvus, H. labyrinthica* and *pulchella,* and *Pupa contracta;* but they are more partial to swampy situations, and with other small species are found in great numbers in certain places farther back in the country. Just one dead shell of *nitidus* has been taken—near the water, and it would seem to be a splendid locality for them. The *albolabris* is worthy of special mention on account of the superior size to which they attain: very seldom are they less than 30 mill. in diameter, while one shell measures 36. The *H. palliata* also are very perfect.

From the lower end of these woods to the 'Vly' is but a short distance; a long narrow strip of woodland lies on the north side of the causeway and forms the entire south shore of the cove. Here the conditions are much different; the ground is not over a foot or two above the high tides, and portions of it are occasionally inundated. The soil is of rich black mould with clay substratum, and has produced a dense growth of trees, principally elm; and a luxurious, almost tropical, undergrowth of shrubbery, ferns, and weeds.

Here lives and flourishes a colony of *Succinea obliqua* that is peculiarly interesting. During the warm months, May, June, July and August, they are wonderfully abundant. After the rains they are swarming over everything; feeding on the decaying rubbish, crawling on the weeds and bushes, going up the trunks of trees, and disporting themselves generally as if they really enjoyed their existence. Sometimes I have observed eighteen or twenty large fellows gathered around the foot of a tree as if on the point of a forward march of ascension. They never go very high however; I have not noticed them beyond five or six feet from the ground. Nor do they confine their attention to the woods; for in an adjacent large meadow many of them may be found traveling in the deep grass, some as much as a third of a mile away on the hillsides. So congenial are all the conditions surrounding them that they grow to surprising proportions; the best shells average 24 to 25 millimeters, often exceeding this. I have recently obtained one that is 28 mill. long. Mr. Pilsbry, to whom I sent a few specimens, says of them, "they are simply phenomenal in size." Mr. Binney tells me one rarely meets such large ones. The greatest length he mentions in his Manual of American Land Shells is 25 mill. Toward the latter part of summer the older ones die off rapidly, and late in the fall very few of them can be seen—but some of course survive the winter, while plenty of young will be left in the field for another year, which hibernate so carefully that one is amazed when spring opens to find such armies of them.

Living along with *Succineœ* are *H. thyroides* and *alternata;* shells of the former quite pretty, some of them delicate pink color, and a number of specimens are encircled with two or three bands of white, seemingly eroded. *Macrocylis concava* and *Zonites fulvus* also occur. *Pupœ* are scarce; I have only seen a few *contracta* and *pentodon*. In the wettest parts of the woods, in the moss, great numbers of *Pomatiopsis lapidaria* can be gathered; also *Carychium exiguum;* and in the cove and river in the near vicinity are twenty or more species of fresh water shells, many of them of excellent quality.

H. thyroides.

During the early part of the present winter, as frosty days were quite the exception, I visited "Almont" frequently for collecting, all of them delightfully successsful trips. Have gleaned much of interest regarding the hibernation of the different snails there found. Here are my notes for the 7th of January this year:

Selenites concava.

"Particularly numerous at this time are *H. palliata*, though not so easily found as in summer. They are invariably closed with the epiphragm, lying aperture upward, looking very pretty when first exposed to the light, their pearly white lips contrasting beautifully with the dark epidermis. Old bark nests seem to be a favorite place for them to congregate for winter. Sometimes they will be found singly, often five or six grouped together; and at times as many as twenty or thirty distributed about a single little vicinity. A situation of this sort is often chosen by *H. monodon (fraterna)*; this species can thus be found to the extent of twenty or more individuals in a cluster wintering along with *H. palliata*. Once in a while the collector is pleased by the finding of a large *Zonites fuliginosus*

Z. fuliginosus.

buried his whole depth in the ground, and nothing visible save the membranous covering over the aperture. *H. albolabris*, usually so plentiful in the warm season is now apparently very scarce; not over a half dozen live ones found this winter, and they were among the leaves, partially imbedded. In another wood near here the boys while raking leaves late last fall obtained for me about one hundred specimens hibernating in the same way. *H. thyroides* at this time is occasionally gotten here and more especially at the 'Vly,' mostly buried 'n the earth. In a few instances I find the animal out and crawling, observed them to-day, and on the 26th of December. A cluster of very well-developed *fulvus* was obtained on the latter date under stones near tide water. A goodly quantity of *S. ovalis* was gathered a while ago, among and attached to broken rushes between the tides (dormant); but their number has greatly decreased since last summer."

Thus the region is more than doubly interesting to the conchologist, and it is one of the fields to which I have given considerable attention.

Nomenclature and Check-List
—OF—
AMERICAN LAND SHELLS.

A list of United States Land Shells, complete to Sept., 1889, containing the species and varieties with their habitats, prefaced by a brief discussion of their nomenclature.

Collectors interested in American Shells will find this a useful list for arranging and naming their collections, checking desiderata in making exchanges, etc.

20 pages, Octavo, Price, single copies - 10 cts.
"   "   " per dozen - 50 cts.

Vol. III.   NOVEMBER, 1889.   No. 7.

COLLECTING LAND SHELLS IN SOUTHERN CALIFORNIA.

BY EDWARD W. ROPER.

To a farm boy from Massachusetts and to an amateur conchologist who had collected shells in Jamaica, the dry hills behind San Diego must have seemed a comparatively desolate ground for land snails. When 31 years old, the young newspaper editor, Edward Warren Roper, briefly visited San Diego, California. He submitted a short article to The Nautilus *in 1889. He recorded three of the locally most common land species:* **Micrarionta stearnsiana** *Gabb, a ¾-inch snail named after Robert E. C. Stearns in 1867; the glossy-black* **Glyptostoma newberryanum** *named by W. G. Binney in 1858 after the distinguished geologist, author and physician, John Strong Newberry (1822-1892); and "***Succinea oregonensis *Lea," a lost species. Roper's specimens were probably the common* **Catinella rehderi** *(Pilsbry 1948).*

"Look where you step" is a good rule to follow in any country, but it is absolutely essential in San Diego county, for two reasons. First, because it is very important, if there is a rattlesnake in your path, to see him before treading upon him. Secondly, because if you carelessly step on the little round cactus so common in this region, the spines, if they do not puncture the sole of your shoe, will penetrate the upper leather more surely than needles. In the eyes of an eastern collector, accustomed to look for land shells in moist, shady places, it is not a promising country. There are no woods, except on the mountains, and few streams of water around whose banks mollusks might be expected. Yet there are shells all around.

Find a cactus that is dead, and turn over its fallen leaves with a stout stick. Like the watermelon, a cactus seems to carry its own water, and under this moist, decaying mass the little Pupas may be found, and Helix Stearnsiana Gabb takes shelter from the sun. The night dews are heavy, and doubtless when darkness falls, the snails emerge from their hiding places, and browse around for food.

Another favorite collecting ground is a pile of loose rocks; if on the south side of a hill, where the sun beats hottest, so much the better. Turn over every stone until the damp earth is reached, and your eyes will be gladdened by the sight of the elegant dark brown shiny Glyptostoma Newberryana W. G. B. If the rocks are in the midst of shrubbery and herbage, the large beautifully banded Arianta tudiculata Binn. is likely to be found. Very rarely do any of these shells live on the shaded northern slopes, doubtless because where the ground is less heated during the day, less moisture is condensed at night. In this country, then, the collector truly earns his prizes by the sweat of his brow.

One other land shell is the Succinea Oregonensis Lea, of a reddish golden hue, found on the weedy river banks, and living only a little less in the water than its frequent companions Limnæa Adelinæ Tryon, and Physa Gabbii Tryon. These are the common shells of the open country, although far from numerous in individuals, when one considers the hours of diligent labor necessary to procure a reasonable number.

GENERAL NOTES.

"SLUGS" AS MEDICINE.—While in Port Antonio, Jamaica, last March, I collected some *Veronicella sloanei Cuv.*, and having nothing to put them in, wrapped them in paper and left them on a table in my room at the hotel. During my absence they escaped and began crawling around, much to the disgust of the colored chambermaid who happened in about that time. On my return she filed a vigorous protest against the "nawsty things," and wanted to know what I intended to do with them. She then informed me that they were good for all forms of lung trouble and asthma. They are used as follows: Take a green cocoanut, cut off the end, and drop a good sized "slug" into the milk, in which it will dissolve. The milk is then drunk and is a "sure cure for asthma."

It would be interesting to know whether this is a survival of the old European belief in the efficacy of the slime of "slugs" in pulmonary troubles, carried to the island by the early English settlers, or whether it is a part of the African pharmacopœia introduced with the slaves.—GEO. H. CLAPP.

Vol. XIII. APRIL, 1900. No. 12.

SOME NOTES ON THE LAND SHELLS OF WESTERN FLORIDA.

BY C. W. JOHNSON.

This short article, one of 219 contributed by Charles W. Johnson, is included not only because of its interesting natural history of terrestrial mollusks, but because it demonstrates an almost universal urge on the part of so many naturalists to collect at the slightest opportunity. Johnson was on a fossil-hunting trip for the Academy of Natural Sciences of Philadelphia in 1900, where he was Honorary Curator of Tertiary mollusks, and for the Wagner Free Institute of Philadelphia, where he was Chief Curator from 1888 to 1903. Although mainly a marine mollusk man and an expert in dipterous flies, he collected land snails near the fossil outcrops at Jackson's Bluff, and also while waiting both for the steamboat at Blountstown on the Apalachicola River and for the train to Marianna, Florida. He says "visions of a new species or variety formed an active stimulant; for I felt sure that Hemphill (of California) Ferriss (of Indiana), and Sargent (of Alabama) had not been there."

A fuller account of this beloved man, after whom the mollusk journal Johnsonia *was named, is given in our chapter on marine collecting along the Atlantic Coast.*

The following notes on the land mollusca are based on a few obtained incidentally while collecting fossils in Western Florida during the latter half of February and the first week in March.

These notes give a more southern and western distribution for a number of species than has heretofore been recorded.

The more southern distribution is undoubtedly due to the direct southerly course of all the rivers, which during freshets carry down great quantities of drift-wood to which a number of the land shells usually cling for preservation. A more western range for a number of the eastern species would be expected, and more thorough and extended researches would probably show a much greater distribution westward.

In the woods just east of Tallahassee, among the leaves around the foot of some large magnolias and oaks, a number of *Polygyra pustula* and *P. hopetonensis* and a few *Omphalina lævigata* and *Strobilops labyrinthica* were found. Near by in an old decayed log were found *Glandina truncata* (young), *Vitrea indentata*, *Zonitoides arboreus*, *Z. milium* and *Philomycus carolinensis*.

At Jackson's Bluff on the Ocklocknee river, 24 miles west of Tallahassee, is a fine exposure of the Chesapeake miocene. Here a few favorable logs and stones were hastily turned over; under the limestone was found *Helicina orbiculata* and *Glandina truncata*, while from the logs were taken *Omphalina lævigata*, *Gastrodonta suppressa*, a form in which the umbilicus is but slightly perforate, *G. demissa*, *Vitrea indentata*, *Helicodiscus lineatus* and *Polygyra inflecta*; for the latter species this is a more southern locality than has previously been given.

Two miles below Jackson's Bluff is Larkin's Bluff; under some boards and wood near the Bluff only *Polygyra hopetonensis* was found; this is the most western locality from which I obtained this species.

About half a mile below Bailey's Ferry, on the west side of the Chipola river, 11 miles west of Blountstown, is the farm of Mr. J. P. McClellan; here the Chipola bed comes to the surface and the shells are ploughed out in the field. After obtaining a fine lot of the Chipola fossils and several boxes of the marl from which the clay and sand had been washed through a seive, I turned over an old log, just as I was leaving, and found *Gastrodonta intertexta*, the strongly carinated form, but with the usual internal callus. *G. demissa*, the most southern locality from which this species has been recorded. *Polygyra appressa* var. *perigrapta*, formerly recorded only from the mountainous portions of Tennessee and adjacent States, *P. inflecta*, and *P. pustula*, which has not before been reported west of Cedar Keys. In crossing the field near by I found an immature specimen of *P. albolabris*.

While waiting for the steamboat at Blountstown a short stroll was taken through the woods; a search beneath the oak logs disclosed a number of *Polygyra fallax*. It seemed strange how these were confined exclusively to the oak; numerous pine logs were turned over, close by the oak, and all conditions seemed equally favorable, but not a single shell was obtained. *P. fallax* has not to my knowledge been recorded south of northern Georgia. Under the bark of logs, in the drift along the Apalachicola river, was the ever present *Zonitoides arboreus*.

As the steamboat did not connect with the east-bound train, I was obliged to go to Marianna for the night. I had noticed from the car window the week before an outcrop of limestone at the railroad bridge across the Chipola river, one mile east of town, that I wanted very much to examine, so before train time, the next morning, I made a grand rush for the river. The nummulitic limestone contained but one mollusk, *Pecten perplanus*, but what it lacked paleontologically, it made up malacologically in furnishing a suitable environment for numerous species of snails. A glance showed it to be an ideal collecting ground; limestone, moisture, a varied vegetation, a cave and an old quarry with moss-covered rocks in all directions, is just what the snails want, and visions of a new species or variety formed an active stimulant; for I felt sure that Hemphill, Ferriss or Sargent had not been there. But alas, while the snails were thick, a *nov. sp.* was not to be found by " dis chile." Ferriss *would* no doubt have found one, for I still believe it's there. *Pyramidula alternata* was very abundant, a coarsely sculptured and beautifully marked form, among which I found a sinistral specimen. *P. perspectiva* was also plentiful; neither of these have previously been recorded from Florida. Among the leaves in front of the cave were numerous fine specimens of *Gastrodonta demissa*, the majority of which are slightly more depressed than the typical form. *Omphalina laevigata* chiefly frequented an old log, while *Helicina orbiculata* were found among the rocks in the drier portions of the quarry. A few specimens of the following species were also obtained: *Glandina truncata, Zonitoides arboreus, Vitrea indentata, Strobilops labyrinthica, Bifidaria armifera, Polygyra inflecta, P. appressa* var. *perigrapta,* and *P. stenotrema*. The latter species have not before been recorded from Florida. In the river drift near the bridge were numerous specimens of *Polygyra auriformis* and a few *Succinea luteola*. As this drift was not the direct wash of the river, but was formed by the water backing up over the low ground along the railroad, I am inclined to think that the two species could be found living among the grass and sedge along the high-water mark.

JAMES W. QUEEN & CO.,
1010 CHESTNUT ST., PHILADELPHIA.

Microscopes and Supplies, Magnifying Glasses, Insect Pins, Sheet Cork, Plant Presses, Collecting Cases, Etc.

Send for Complete CATALOGUE **B.**, also Sample Copy of the MICROSCOPICAL BULLETIN.

ACHATINELLA HUNTING IN NORTHWESTERN OAHU.

The shiny, agatelike tree snails of Hawaii, belonging to the genera Achatinella and Amastra, have been collectors' items ever since Captain James Cook had brought back specimens to London in the late 1700's. There were many famous American collectors of Achatinellidae, the early ones being Dr. Wesley Newcomb of Albany, New York (1818-1892), William H. Pease, a land surveyor (1824-1871), Joseph S. Emerson, a civil engineer (1843-1930) and David D. Baldwin, a teacher and sugar plantation manager (1831-1912). Later came Joseph Gouvei, Irwin Spalding, d'Alte Welch and Dr. C. Montague Cooke. With each passing generation the shells became scarcer and much harder to find, until today only a few of the many hundreds of species still survive the onslaught of land developers.

We take the liberty of printing extracts from a letter received some time ago from Mr. Irwin Spalding of Honolulu, in explanation of the interesting photograph of living Achatinellas reproduced on plate II. As a general rule, these snails are found "sleeping" by day, on the under side of a leaf as in the picture, under loose bark, or in a knot hole. They are doubtless active chiefly by night.

Those who have used the monograph in the *Manual of Conchology* know that many species and color-races once abundant are now rare, some doubtless extinct. Dr. Newcomb and Mr. Gulick collected fine tree-shells in quantity where forests are now but a tradition, and their shells are often of color-patterns strange to the modern collector. It is most gratifying to learn that some of these long-lost species are being turned up at higher levels. Mr. Spalding writes as follows:

"So many good things have come my way along the land-shell line these last two months that I really don't know how to begin and tell you all about them. To start in with, I spent my three weeks vacation this year collecting on Oahu, started in at Opaeula and worked around through Waimea, Pupukea, Waialee, Kahuku, Leie, finally landing up at Hauula. Only 2367 *Achatinella* and *Amastra* for the trip, but here is the best of it all, found four of the supposed extinct species, *A. bulimoides; A. emersoni;* typical old-time, banded mottled *A. curta,* and last but not least, *A. ——?;* the latter to be seen in the accompanying photograph of six fine adults on a leaf [Plate II]. I will give you six guesses, if you guess their identity I will send you six. As I have hinted more or less where it comes from, I suppose you have guessed correctly at least once out of the six trials, so by bearer you have your shells [they are *Achatinelli elegans* Nc., from Hauula, long supposed to be extinct]. What do you think of that for a find? The first trip netted me 40, second 13, and the last 134, each trip representing as many different ridges. I consider this one of the best land-shell finds of the last dozen years or so.

"*A. bulimoides* is as good as extinct. The first day netted me 7, second day but one. They look very much, as Wilder says, like a reversed *rosea*. The two other species we struck in small colonies, collecting probably a hundred of each.

"I was greatly disappointed in the Waialee district, finding none of the old-timers reported from that section.

"In the fossil bed at Kahuku I found what Montague Cooke claims is a new species of *Amastra*, a form between *Leptachatina* and *Amastra*, small and cylindrical.

"I am glad that you have come to the conclusion that there are too many *Pterodiscus* named from Oahu. It was only a couple of weeks ago that I struck a locality west of Palikea in the Waianae Mts. Not the so-called Palikea where Thaanum found his *heliciformis*, according to the Manual. His locality is Green Peak, marked erroneously on maps as Palikea,—the same place where I found this species some six years ago. Palikea is the high peak northwest of Pohakea gap. Anyway I collected well up to a hundred of a species, samples of which I also send."

JAMES H. FERRISS
November 18, 1849–March 17, 1926

James H. Ferriss has been called the "Mark Twain" of the American conchological world, for he had a delightful and humorous style, was a newspaper editor, and lived about the time of Samuel L. Clemens of Hannibal, Missouri. Were it not for Ferriss, Henry A. Pilsbry in Philadelphia would not have been as active in describing new American land species nor have gone collecting in the southwest and the Great Smokeys of Tennessee. It is possible that Pilsbry's monumental Land Mollusca of North America *might not have been started or even have been completed satisfactorily had it not been for the collecting ventures of Ferriss and his devoted conchological companions.*

With the railroads newly spanning America and the automobile just coming in, the period from 1880 to 1914 was a time of conchological pioneering west of the Mississippi. A shell "gold rush" was on, with Ferriss, George H. Clapp of Pittsburgh, Bryant Walker of Detroit, H. E. Sargent, and Henry Pilsbry in the lead.

Pilsbry's biography of Ferriss properly sets the stage for the charming series of adventures published in The Nautilus *from 1899 to 1924 by this foremost of American land-shell collectors.*

James H. Ferriss, the foremost of American land-shell collectors, was born in Kendall Township, Kendall Co., Illinois. His father, William H. Ferriss, was a native of New York State. His mother, Eliza M. Brown, was born in Lowville, Erie Co., Pennsylvania. On the maternal side the family had long lived in or near Erie, Pa., where his grandmother, Adeline Sloane, was born in 1811

In the fifties and sixties educational facilities in rural Illinois were not very good, and Ferriss' formal schooling did not go beyond the high school stage. He had a taste for reading and a retentive memory; in addition to these, he had a combination of qualities rarely found together in so high a degree: an intense interest in and sympathy for his fellow man, and a great love of nature. He was equally at home and happy in the turmoil of a hot campaign against the Joliet saloon and underworld forces, and in a lonely desert camp. That Ferriss did not go further in the more technical side of natural history was merely because life is too short; when at home, the demands of a large daily paper and innumerable details of park and other civic affairs absorbed his time.

As a boy, Ferriss at first worked with his father in the business of buying cattle for the Chicago market, but in 1869, at the age of 19, decided to start out for himself, taking up a claim in southeastern Kansas. Until 1872 he was by turns farmer, freighter and storekeeper. It was the impression made at this time of the evil effects of drink in a pioneer community which made him a staunch advocate of temperance throughout his life.

In 1872 Joliet had a great access of population due to the establishment of steel mills. A "boom" was on. Tired of frontier life, Ferriss returned, obtaining a position as reporter on a Joliet paper. His knowledge of cattle and trade soon made his "market page" a prominent feature. Two years later he started a paper of his own at Yorkville, county seat of Kendall Co., Ill., but returned to Joliet in 1877 to take the management of the *Phoenix*. A few months later, with H. E. Baldwin and R. W. Nelson, he purchased the *Joliet News*. There was scarcely any real money in the Middle West at this time. The three young men, without capital or backing, hardly knew from one day to the next where to get money to issue the paper. One of Jim's fiery editorials offended a local political boss, and the editors of the *News* were jailed. The paper missed an issue, but release came the next day. The *News* was in the right, and became popular over night. New subscriptions and advertisements poured in, and the three proprietors did not miss any more meals.

From this time Ferriss was continuously editor of the *News* except between 1880 and 1882, when he edited a Maine paper backed by Neal Dow, the noted prohibitionist of those days. Finding the conditions irksome, he returned to Joliet, where the *News* thereafter was conducted by him as editor and Baldwin as business manager. In civic affairs the *News* stood for the best man, regardless of party, for better schools and a cleaner city. It was the terror of grafters and saloon politicians. In National affairs it supported the Populist party—an expression of the dissatisfaction of the Middle West with existing conditions. Loans contracted in the expansive years now had to be paid in a contracted currency. It was the era of unfair railroad tariffs and unrestricted monopolies. With the passing of these conditions the party dissolved. Ferriss was National Chairman in the Convention of 1904. In 1915 he retired from active newspaper work. The *News* was merged with the *Herald*, becoming the *Joilet Herald-News*.

As a youth Ferriss was interested in the natural sciences, at first collecting trilobites and other fossils of the local Paleozoic. His first conchological discovery, I believe, was *Unio superiorensis* Marsh, which he found on one of several fishing trips to the north shore of Lake Superior, in the early nineties. Gradually the collection of mollusks and ferns replaced fishing as a vacation pastime. About 1896 he discovered the charms of the southern Alleghanies. "Surely it is an enchanted land," he wrote, "for I am homesick until I return." For the next six years he made annual visits to this lovely mountain region, in 1898 with George H. Clapp, in 1899 with Bryant Walker, Clapp, H. E. Sargent and myself. I suppose we will never forget this glorious trip. We felt that the combination of ravishing scenery, new snails and congenial companionship could never be surpassed.[1] In the following year Ferriss and Walker explored the mountains northward. Many new and fine land

---

[1] NAUTILUS, XIV. Proc. A. N. S. Phila. 1900, 1902. It was in this region of East Tennessee and Western North Carolina that Miss Law had found *Polygyra chilhoweensis*, and Mrs. Andrews *P. andrewsæ* and *Vitrinizonites latissimus*. Ferriss thought that if women could discover such splendid species, a man ought to find one "as big as a tincup, with spines."

shells rewarded Ferriss' work in this region. Among malacologists, his name will always be associated with it, together with those of Rugel, Mrs. George Andrews and Miss Annie M. Law.

In February of 1899 and 1900 Ferriss made hasty trips into southwestern Arkansas, and early in 1901 a further exploration more northward and in Indian Territory (now Oklahoma). In March and April, 1903, Ferriss and I, starting in southwestern Missouri, hunted in other places in Arkansas and Indian Territory, thence into Texas, as far as the Devil's River and the spectacular canyon at the mouth of the Pecos. Among the new snails turned up in these several trips were such elegant helices as *Polygyra binneyana*, *indianorum*, *pilsbryi* and *unicifera*, the rare *P. kiowaensis* Simpson, and in Western Texas the genera *Daudebardiella* and *Cochliopa*, new to our fauna. *Lampsilis simpsoni* Ferr. was also from one of these trips.[1]

February, 1902, was, I believe, the date of Ferriss' first Arizona collecting trip. It was a hasty one, mainly in quest of rare ferns, as he was trying to complete his collection of living U. S. ferns for the Joliet park. Land shells had second place, but he picked up enough to show the richness of the fauna, up to that time scarcely known. In 1904 snails had first place, and fine species of *Sonorella*, *Ashmunella* and *Oreohelix* were taken in the Huachuca and Chiricahua ranges. He described *Ashmunella walkeri* (Florida Mts.) and *Oreohelix clappi* at this time. The 1906 trip was with me, first in the Grand Canyon region, then south in the Florida and Chiricahua Mountains. Early December found us in the magnificent forest around the head of Cave Creek, the highest part of the range. Bivouacking at 10,000 ft., we woke one morning to find the world white with snow, and waded down to camp in Cave Creek through two feet of snow—a long day's tramp. In 1907 and 1908 Ferriss again visited the range, getting a few additional species in the southern canyons; in 1907 with L. E. Daniels. The land snail faunas of the Huachucas and Chiricahuas were described in 1909 and 1910.[2] No other ranges have supplied such remarkable assemblages of Sonorellas, Ashmunellas, Oreohelices and Holospiras.

In 1909, in company with L. E. Daniels, a trip was made across the Grand Canyon, on the Kaibab and Powell Plateaus. *Oreohelix strigosa depressa* in great abundance and variety was found, also the rare *Succinea hawkinsi* Baird. The results were published, together with those of our trip of 1906, in 1911.

In the expedition of 1910, L. E. Daniels and I joined. We camped in many ranges not before explored for shells.: the Santa Ritas, Baboquivaris, Dragoons and others. *Sonorella ferrissi*, *S. walkeri*, *Holospira danielsi* and *Oreohelix ferrissi* were part of the loot. There was something new about every day. Daniels introduced hot lemonade as a pick-me-up after a hard day's shelling. Ferriss would take his tin cup full before the final ingredient was poured in.

In the limits of my space, all of the trips in Arizona and New Mexico cannot be described, but the chief ones may be mentioned.

1913. Catalina, Rincon, Tortillita and other ranges; afterward with Frank Cole "a splendid guide and biscuit maker" to the White and Blue Mountains, on the Gila headwaters.

1914. With Daniels, into the Mogollons, N. M., where an entirely new set of Ashmunellas turned up.[1]

1915. The Black Range, N. M., with the writer. From a series of camps along the ridge at 8–10,000 ft., and in magnificent forest, we collected such fine Ashmunellas and Oreohelices as *A. cockerelli*, *O. pilsbryi* and many others.[2] This was the last of the trips by wagon and pack train. Ferriss bought a Ford.

In 1917, 1918 and 1919 the ranges west of Tucson to Ajo, and those along the Mexican border, Tumacácori to the Whetstones, etc., were explored, partly with Frank Cole, the best of guides, partly with Hinkley and Camp.[3] *Bulimulus nigromontanus* Dall was one of the notable finds, the genus being new to Arizona.

The summers of 1916–17 were spent with the Sierra Club and in other travels in California.[4] In 1919 with a party in northeastern Arizona, an arid region, but of great interest for its wonderfully preserved cliff dwellings, natural bridges and Mt. Navajo.[5]

Early in 1921 a camping trip in the ranges around Death Valley was carried on in company with E. P. and Mrs. Chace. *Micrarionta* was abundant and varied. Later, with Prof. Edwin E. Hand and Dr. W. T. Miller, he went northward through central Arizona.

In the autumn of 1922 Ferriss and I took the field in historic Santa Fé. The route was down the Rio Grande to the Organ Mountains, across the White Sands and desert to the Sacramentos and north to Nogal Peak, then into the Guadalupe range in southeastern New Mexico and Texas, and south to the Chisos Mountains in the Big Bend of the Rio Grande. The results of this trip will soon be published; there is a great deal of interest, since much of the ground was new. Early in 1924 Ferriss again visited western Texas.[1] His last trip, in the spring of 1925, was in the same region, but cut short by ill-health.

In later years Ferriss became increasingly interested in cacti, collecting great numbers for use in the beautiful monographic work of Dr. J. N. Rose, and for his own collection in a large greenhouse built for their cultivation in the West Park, Joliet. To spin along over a rocky desert with Ferriss at the wheel and at the same time looking sharply for wayside cacti was an education for the nerves. Very few large stones were missed. On a long trip the only place in the car free of cacti would be the bed-roll.

Ferriss was an ideal camping companion. He loved life in the open, had great endurance, an unfailing optimism, and an exhaustless store of entertaining talk. He was at his best in camp, and drew out the best in others. The prospector or cattleman who chanced to drop into camp often stayed swapping reminiscences around the fire far into the night—tales of the Indian times, of Apache Kid, Cochise Stronghold (where we camped), and of course, of the search for lost mines. Nearly everywhere Ferriss found somebody he knew, or who knew a friend of his, and he was heartily welcome in hundreds of places throughout the southwest.

As a collector he has probably never been surpassed. Ferriss found more new land shells than any American since the time of Thomas Say; and he generously shared his finds with those most interested.

The portrait now given is an excellent likeness of Ferriss as I first knew him. In later years his face, always strong and fine-featured, became deeply lined and bronzed by exposure.

Ferriss had no children. His wife survives him.—H. A. P.

THE GREAT SMOKY MOUNTAINS.

BY JAMES H. FERRISS.

There was a general round-up of the snails in the Smokies last summer. When the roll of diggers was called at Cades Cove, Dr. H. A. Pilsbry answered to his name, and so did Geo. H. Clapp, of Pittsburg, Bryant Walker, of Detroit, Prof. H. A. Sargent, of Ann Arbor, and I did too. Prof. A. G. Wetherby and Mrs. M. L. Andrews intended to be with our party until the very last moment. The year before, I made the trip as far as Mirey Ridge with Mr. Clapp. With, this exception it was my first excursion in company with up-to-date scientists. I have made four trips to the Smoky Mountains and expect to go again this year. On two occasions short stops were made at Burnside, Kentucky, on the Cumberland; at Oakdale, Tennessee, on the Emery; Lookout Mountain, at Chattanooga, and a side trip to the Little Tennessee, at Caringer post-office, or Talassee Ford, and one trip was made into the Unaka range. The Smoky Mountains on the north of the Little Tennessee and the Unaka range on the south (not the Unakas near Roan Mountain), form the boundary between Tennessee and North Carolina.

The readers of the NAUTILUS, I am sure, will be pleased to know something of this party. Briefly in ages, its members ran from 35 to 50; at least I am that high, but they are boys still, and can climb more trees and wade streams worse than ever. Mr. Walker, an attorney, and Mr. Clapp, a business man, I think the handsomest members of the party; and their dispositions, their patience, their interest in the comfort of others really approach the domain of the angels, and when Mr. Blair, our mountain host, was with the party it made three of them. Mr. Clapp can suffer more and complain less than any entirely earthly being. When lame enough to put an ordinary man in a hospital he will sprinkle on a little talcum powder, keep up with the procession and never say a word. Mr. Walker did not sleep the night after our party separated because Sargent and I were out on the mountains without blankets, and the heathen, the two of us, at that very time were as near the happy hunting grounds, both in altitude and spirit, as we may ever be; with a bed of dry moss and a roaring fire at our feet, we slept sweetly as doves, under a massive balsam in the prettiest park I ever saw in the mountains. The next morning we got over 80 *Polygyra Ferrissi* each, and three were albinos.

For industry, zeal and business (shell business), Sargent and Pilsbry are not to be excelled. Sargent always hunts longer and gets more than any other, and Pilsbry, after a hard day's digging, was ready to clean up my catch any time I would bake biscuit. Not one was a believer in ghosts. It was the most sensible, kindly, lovable collection possible. A sad day came when the company sepparated. Dr. Pilsbry then borrowed soda of a herder and attempted to bake his own biscuit. He did not have any sour milk, and I think that yellow spot remains in the camp site to-day, a wonder to passing herders and a puzzle to those practical mountain scientists who condense their bulky corn crop into convenient form for transportation in jugs.

Cade's Cove, in Blount county, Tennessee, lying at the base of the Smokies, is 1,700 feet above the sea. It is six miles long, in some places two in width, and out of this valley are many other deep coves running up to the top of Boat and Rich mountains, 3,500 feet above the sea. This valley has been searched more than any we have visited. But last year we found four more kinds, and one of those a new variety. The soil is so fertile in shells, like the sea coast of Florida it will be good ground for many years.

With mountain friends, camp dunnage and mules, we left the settlement soon as possible. There was much rain, and the puncheons in the herder's cabin where we slept the first two nights were very hard, but it was a light-hearted company. There were plenty of snails, and school children were never more delighted or delightful. The pleasant days we climbed the mountain sides, when Mr. Pilsbry and company talked snails, geology, botany and fungi, is a memory will long live pleasantly with your humble author.

Thunderhead is 5,500 feet according to the government maps, and it rains there every week in my experience and it is more stormswept than many of the higher peaks. The beech trees and buckeyes are mere scrubs. Blockhouse mountain, of the same height, Coontown, Russell's field and other good coves were hunted over from the first camp. Then we moved along the backbone of the range to Clingman's Dome, some 15 miles farther, passing Briar

Knob, the Derricks, Mirey Ridge, Siler's Bald and the Balsam, all over a mile high, and good collecting ground.

Clingman's Dome is 6,600 feet high, covered with balsam fir, and the sphagnum is so deep walking is like tramping on a spring mattress, and very tiresome. When away from a well-beaten trail it is difficult to walk a mile in less than an hour or an hour and a half. Many of the rocks were large as houses, and when we went under for rare shells we carried candles. These feed on the microscopic fungi, I suspect, growing upon the roof, and they seemed to select a roof nearly level. One of the *P. ferrissi* at a time is the rule, but on Andrews Bald, afterwards, we sometimes found as many as eight on one roof. Occasionally *P. clarkii, andrewsæ altivaga, depilata,* or a *Gastrodonta lamillidens* or *clappii*, would be found on the same roof, but not often.

Bidding the remainder of our party and the mules farewell, as our vacation was longer, Prof. Sargent and I, with a couple of mountain friends, carrying our camp outfit upon our backs, parted company from Pilsbry, Walker and Clapp, and made a trip to Andrews Bald (5,900 feet) from Clingman, though we really started out for Mt. Collins, some 600 feet higher. On Andrews, besides *ferrissi*, we found our finest red *andrewsæ altivaga*, banded with a still darker band.

The next day we retraced our steps over Clingman and the Balsam to Siler's Bald, where we took the Welsh Bald trail and continued in a southwesterly direction in North Carolina for the next three weeks, with the exception of the two last days. Sometimes we were on the trail all day, while on other days we went only a mile or two. Sometimes we stayed several days in one place. The weather man furnished his best, and only twice were we compelled to build bark shelters to keep us dry.

On Welsh Bald, at an altitude of 5,000 feet, we first found the new variety of *Polygyra edwardsii*, and from a little spring that oozed out from near the top, we found *Pisidium roperi* Sterki. Sargent found this in Minnesota and I had found it in a small pool near Joliet, but the shell is still rare.

We descended to Chambers' Creek one hot afternoon, where it was only 1,500 above the sea. It was a tough slide and both of our mountain friends were sick before starting. From there Sargent made a side trip by rail to Hayesville, N. C., and I first found *Poly.*

monodon cincta. And then and afterwards they were mostly dead and found around the basswood and buckeye trees. After a few days' rest, we crossed over to Tuskegee Creek, and in Ramp Cove, on the Tuskeegee side of the Yellow Creek Mountains, we first found *Gastrodonta Walkeri* Pilsbry, a new species. It was in company with *significans*. These mountains run up about 4,000 feet, with soil on the slopes rich as a garden.

Passing down Yellow Creek, between the Cheowah and Yellow Creek Mountains, we loaded up with green corn, sweet potatoes and other good things, as the valley is settled. Here we discovered that *Poly. christyi* has a great fondness for the shrub called poison hemlock. The streams were swift and rocky. We found no clams and very few univalves.

At Cheowah river we were down to 1,500 feet again. Hangover and Mount Hayo, in the Unaka range, 5,200 feet, overlook the ford, and the trail we took to these peaks was up a dry pine ridge, steep as the roof of a house, and for the first time in our trip, good drinking water was a little scarce. It took us until 3 o'clock in the afternoon to get up, and all were sore and some were cross.

Every day brought new delights. One afternoon, on Bob Stratton's Bald, 5,400 feet (there is another peak a few miles away called John Stratton's Bald), near Hayo, we found over 200 *G. lamellidens*. We found these in company with *Helicodiscus lineatus*, and *Vitrea carolinensis*, by turning over slabs of stone that lay on top of the ground, and there were sometimes a half a dozen under one stone. The general rule is one *lamellidens* to a dozen or two rocks. The next day, at Glen Cove, a couple of miles lower down the range, we found 130 *Poly. chilhoweensis*. Back on the Little Tennessee river again at Talassee ford, we again found *Gastro. walkeri* at a point less than 1,000 feet above the sea—the lowest point in our trip. One of the mules and a good walker came to our rescue at Talassee ford and we returned to Cade's Cove, 25 miles in a day. In all we traveled about 150 miles, as measured in a straight line, besides our side trips.

There is much land for the snail hunter here. From the highest peaks we could see mountains 125 miles distant, and it was all mountains as far as we could see in three directions, and over much of this roughness no specimen hunter has traveled.

THE NAVAJO NATION.

BY JAS. H. FERRISS.

Sixty miles west of the corner post of Arizona, New Mexico, Colorado and Utah, the 1919 summer class in archæology, Arizona University, encamped at the foot of Navajo Mountain. Here is the greatest number of ancient cliff cities and villages and the greatest of known natural bridges. In scenery, colors, heroic size and architecture, it is Grand Canyon in character. Navajo Mountain astraddle the Arizona-Utah line stands on the south rim of the Grand Canyon, a short distance above Marble Canyon and Lee's Ferry.

In reality the region from the Mesa Verde National Park, Colorado, on the east, to the Zion National Park, Virgin River, Arizona, on the west, it is something of a wonder-spot of the world, and all of it astonishing. The greater cliff ruins, Mesa Verde, Keet Seel, Betatakin and many others as interesting; the Monument Park, a plateau of natural pinnacles and steeples, and the Chinle and Canyon de Chelly valleys are along the eastern border. Then westward lie the painted deserts, petrified forests, the Grand Canyon, the Kaibab forest, underground lakes of Kanab, lava cones of Mount Trumbull, Hurricane Fault, Grand Wash, canyons of Virgin River, plains of wild horses and the largest Indian population in the United States still living in the Indian way. Except to the explorers, archæologists, geologists and mineralogists it is the great unknown of America, and the farthest from a railway.

Dr. Byron Cummings, dean of archæology, Arizona University, and his explorer-companion, John Wetherill, post-trader and postmaster at Kayenta, Ariz., have explored and studied conditions here at this eastern border for more than twenty years, and by right of discovery (as in conchology) should have their names attached to the greater number of ruins and bridges, for they have been the first discoverers, scientifically. Herbert E. Gregory[1] for the government has made a thorough geologic survey of this eastern section covering the Navajo nation, some 22,725 miles. Others before Gregory have written and surveyed, but he is the latest and best authority. Col. Roosevelt and his boys, Zane Grey, the Kolb Brothers and other strenuous persons have visited the Rainbow Bridge but not over 150 white people all told have made the journey. Thus to the students in botany, archæology, conchology, entomology and the reptile hunters, it is a field of great promise. The health seeker and tourist will soon follow, and with profit.

The Indian population of the Navajo country as estimated in 1912 was 32,000, of which 30,016 were Navajo, 2,272 Hopis ("Moquoi" is a Navajo nickname for the Hopis), and 200 Piutes and 521 white Indian agents, teachers and traders. North of the San Juan River in Utah and Colorado adjoining is another large reservation of Utes.

In northeastern Arizona sandstone and shale of different periods are the prevailing types, geologically. The Carrizo range and a small country about sixty miles north of Holbrook, Arizona, and a few needles thrust through the desert floor here and there belong to the igneous group. Less than one-tenth is limestone, and in character of little worth to the snail industry.

Vegetation is not so plentiful or varied in character as in the region lying southward to the Mexican border, but much of the material is new to collectors, and some of the species new to science. At an elevation between 6,000 and 7,000 feet juniper (*J. monosperma*) and pinyon (*P. edulis*), and up to 8,500 feet yellow pine (*P. ponderosa*), quaking asp, spruce and oak prevail, with columbines, phlox, aconitum, larkspur in the usual mountain profusion. Ferns are rare.

W. N. Clute, editor of the American Botanist, and the present writer, both of Joliet, Ill., were invited to join the class of 1919. They needed no urging. The good ship Ford, chafing at its Tucson anchorage was in line at Flagstaff July 1. Leaving the Lowell Observatory with the Normal School faculty and several pleasant people, a run was made over to Grand View, on the Grand Canyon, about 70 miles, to organize, get acquainted and make a fresh start. It is one of the best views of the canyon. The auto parties camp here, but the hotel, now owned by W. R. Hearst, is idle. The forest rangers' camp is nearby but otherwise there is no settlement here at present. *Sonorella coloradoensis* (Stearns) was found at the type locality. Scenery, fine air and the yellow-blooming century-plant (*Agave utahensis*) were of particular interest.

Tuba City was our next camp, and it was necessary to return to the San Francisco Mountains, 14 miles from Flagstaff to cross the Painted Desert. This is the fourth time we have passed this range with a peak of 12,794 feet and everlasting snow. Surely Oreohelix is up there in the quaking asps, but no conchologist has made a track so much as on the foothills.

At Cameron (Tanner) crossing was made over the canyon of the Little Colorado on a suspension bridge built by the government. There was but a thread of water in the muddy flats one or two hundred feet below. A little scratching here during the luncheon hour did not turn up anything in the rocks or drift. The road was fairly good and the autos hummed along merrily over wide stretches of black lava, pebbled agate, iron marbles and other geological curiosities, and at other stretches, for a change, painted canyon walls and miles of grotesque windmade statues furnished entertainment. Although delayed four hours in starting we traveled 122 miles and went into camp early that day at Tuba. Purchased from the Mormons, this is now a government city of schools, agency buildings, an agricultural experiment station and a hospital.

Sand is the chief product of Tuba City, but springs are numerous and the fields of grain and the orchards were thrifty. At the boiling spring in our camp a new Pisidium and *Physa humerosa interioris*, n. subsp. were gathered. Also cases of the case-fly and a fair collection of dragon flies. A large scarlet species was the prize.

John Lee, of Mountain Meadow memory, was one of the founders of Tuba. Later he established Lee's Ferry on the Colorado, and later again at Mountain Meadow met his Waterloo. On the rocks of Moenkoppi Wash is the village of Moenkoppi. The homes and stores of these ancient cliff-dwellers were closely inspected by the class, also their fields of corn, orchards and vineyards, hundreds of acres. School diplomas, photographs and three-colored illustrations decorated their walls, and clocks and sewing machines seemed home-like. They are neat housekeepers, hospitable and surely happy.

These so-called Quaker Indians, the Hopis, and also the more or less war-like Navajos, Utes and Piutes, with a few goods bought from the traders—salt, sugar, baking powder and calico—live as they have always lived. They are farmers with fields of grain, alfalfa and vegetables in the low spots of the desert; operators in live stock, manufacturers of blankets, pottery and jewelry. The estimate of 1912 gave this nation 330,000 horses, 33,000 cattle, 1,500,000 sheep. They dress in styles of their own, in dwellings cling to their ancient architecture and keep their blood pure Indian. The Hopi has permanent dwellings, four and five stories high, and perhaps may be the original inventor of the Philadelphia sky-scraper apartment. The Navajo with his solitary and temporary hogan of sticks and mud, the Ute and Piute with tepees of skin or canvas, follow their flock to the herding grounds, all at peace, one with another, really not knowing tribal boundaries. There may be a two-thousand-dollar auto in the front door yard of Mr. Navajo if the ground is that level. The remainder of the family surplus may be invested in government bonds, a banner with a star in gold hanging from a door that is something like a muskrat home, but they make their own moccasins and calico breeches, and some of them still think they can whip the United States. Since we broke camp one of those cockey white men, prospecting for minerals against Indian instructions, was found lying by a water hole on our trail and the signs of his taking-off were Navajo.

The Hopi is a model Indian. He saves his money, never had a quarrel with Uncle Sam, and without government bounties has made his own living. A trader told us that when a Navajo sold him ten dollars worth of wool he traded out the full amount and asked for nine dollars more of credit, but the Hopi left a quarter and took home nine seventy-five in cash.

The Navajo refused to dig for pottery, as the flu had given them a scare; but we liked them and their splendid horses were kindly, well-broken and intelligent. Without a guide or guidance they carried us for miles over naked sandstone where there was not a scratch to mark the trail. Saddles of Navajo make are in good taste and stand up well with the best of the saddler's art.

"Boy, boy," came from a group of smiling Hopis at Tuba, pointing to cavalry pantaloons as the girls climbed into cars. The dogs barked, and with the government veterinarian in his own Ford to lead us, the sand flew over the dunes to Kaibito. Here Wetherill and his horses had been waiting for three days. The cars were stored in the trader's wool house, and two days later we threw down the shovels and anchored those horses and mules in the junipers of Endische Springs at the south foot of Navajo Mountain.

There had been little opportunity for collecting, but while watering our stock at two branches of Navajo Creek the drift was found rich in small shells. The streams contained excellent drinking water and were twenty or thirty feet in width. A large amount of timber had been floated down from the mesas, but we were traveling fast. *Physa humerosa interioris* was found in the stream among the horsetails and water cress.

With the best of water spouting from the rocks, a beautiful view and a delightful climate we settled down into a permanent camp, began to feel acquainted and call each other by front names. These pupils and instructors were a splendid group of uncomplaining pottery diggers. Nearly every western state was represented. Their forebears had been pioneers from Plymouth Rock to California, and thus good sense and the square deal came just natural.

From the southern approach Navajo Mountain is an oblong dome, regular in form, longer east and west, without peaks or precipices, "rising four thousand feet above the flat floor of the Rainbow Plateau, an island in the midst of a sea of water-worn and wind-worn brilliantly colored sandstone," says Gregory. A nearer view and a little travel finds precipices in plenty. In fact so rough were the crags we found but one horse trail to the upper levels, and that ended at War God Springs about half way to the summit. Here at the springs is a fairly level bench in the yellow pine about a mile in width along the southern and eastern slopes. The talus covered with quaking asp largely composed of heavy sandstone blocks is an ideal situation and *Oreohelix* was at home. The summit of about two hundred acres fairly level is heavily clothed in spruce, and over the top, under the precipices are occasional springs that feed the streams crossing the Rainbow trail below. Many fairy bowers, coves and valleys are hidden here for botanists and snail seekers. Here we found a new Phlox (*clutei*); *Oreohelix yavapai cummingsi* n. subsp. and *Gonyodiscus shimeki cockerelli* Pils. were found in their most robust form.

All of the mountain is sandstone, or so near it that shells and their hunters notice no difference. "Cretaceous sandstones cover the top and Jurassic (?) sediments constitute the flanks," to speak authoritively. The sandstone for the whole region is rather variable in character due perhaps to the several binding materials—lime, silicon, iron, manganese, etc. Many specimens were brought to camp, and Prof. Scott's verdict ran to sandstone with an occasional decree favorable to petrified wood.

Navajo Mountain has good soil for snail life, so fertile it is not probable that all the species were gathered. At a spring on the south slope known to us as the Red Rock Spring, *Oreohelix yavapai clutei* n. subsp. was discovered accidentally in the grass and rose bushes. *Succinea avara* was also here in the bogs. Among the rocks of a large canyon west of Endische Springs we found the bones of *Oleohelix yavapai neomexicana* Pils., but found no live ones. This canyon heads in a saddle near the main peak of the mountain and for convenience may be known as Big Pine Canyon until further orders. The north and northeast slopes were not fully explored although three of our party camped at War God Springs the better part of a week. The great rock slides of those slopes probably contain the best snaileries. Four Oreohelix tribes per mountain is a new record for Arizona.

Before returning home Mr. Wetherill and his Indians led the way to the War God Springs and then on foot to the top of the mountain for the view over the San Juan country, and then around the base of the mountain on horseback to the Rainbow Bridge. On the mountain top an *Oreohelix depressa* came to the surface following a shower of rain, in every way almost identical to the shells found by Henderson and Daniels at stations 22 and 23, 1915, near Ogden, Utah. The forest rubbish about the springs was alive with Pupas and Zonitids and Vallonias. A few *Oreohelix yavapai neomexicana* Pils. were in the rock slides.

The outlook from the crest overlooked the Rainbow Bridge, the canyons of the San Juan, other canyons, bridges, caverns, domes, sunlights and shadows, white, brown, and all the reds and all the shades of the amethyst. Also the plateaus beyond the Grand Canyon, the Henry Mountains, 11,410 feet, the Blue, 11,445, Aquarius 10,100, LaSal 12,271. Also the white and black mesas and the Carrizo mountains to the south and east were in view.

The Rainbow Bridge is in the strict rainbow form and with some of its colors. But 30 feet in thickness, with its 309 of altitude and 208 width, in lightness of architecture it seemed something of steel. The average camera does not give an accurate estimate of sharp hillsides and scenery large as this bridge.

It rained a little these evenings, but the bridge kept us dry, and at a camp in Surprise Canyon blankets were spread in wind holes of the cliff. To imitate the swallows, heads and feet were made to peep out a little. At the bridge one of the party imitated the pack rats for a little while and for the first time in his desert experience made a complete collection of fleas. The chute of Zane Grey is an interesting feature of this trail, so narrow it seemed the walls in passing could be touched with either hand, and so high the passage was gloomy. Abduction Cliff and the balanced rock that exterminated the wicked band were true to photographs, one on the trail the other at Navajo Creek, thirty miles away.

In fiction, details in scenery and character should be true to life, though a little latitude may get through of a geographical character. We know Grey's Roaring River, and we camped for weeks at the corral he helped to build for Silver

Mine; we know his Painted Desert, and have struck his trail in so many places we know his details are accurate and well done.

Hon. David Rust, of Kanab, schooled at Leland Stanford University, an editor and twice a member of the Utah legislature, said the only fault "here is that Gray deals in ancient history." Well, so it is with many of us. We do not ask to have witchcraft, intolerance, superstition or any of those disgusting household remedies spread on the records.

David and his son, David Jordan, gave us a pleasant surprise at Endische Springs. They were cousins of mine and it was our first meeting. Our mothers' ancestors, Ezekiel Brown and wife and two sons were kidnapped by the New York Indians and kept in captivity nearly four years. Rust and Ferriss thus inherited their wild ways, and had much in common to talk about. Ferriss all his life, too, because of this family episode, has been tracking New York Indians, especially up and down Wall Street.

A couple of young boys from New York City, taking in the sights from Zion Park to Mesa Verda, were in the care of the Rusts, Arnold W. Kohler, Jr., and Chas. P. Schulzheimer. Though but seventeen they were live wires educationally and went off at the end of a few days with the hearts of us all. They saw a large yellow snail walking up the rocks at Rainbow Bridge, a *Sonorella*, perhaps, but we found only *Succinea avara*, *Physa humerosa interioris* and *Pupilla hebes*.

Loaded with pottery and other material historic, after a few weeks of toil we returned to Kaibeto, assisted by Navajos and their horses. Here we met John Lee, a grandson of the Lee Ferry John, who brought in a report that we were at Navajo Mountain in a starving condition and that the girls had worn out their shoes. The Lees may be a little peculiar, but in a sparsely settled country rumors seem to spring from the ground and spread remarkably fast.

It is a day's journey from Kaibeto to Marsh Pass via Red Lake, by auto or across country by horseback. The Dean and our Navajo friend Leslie made the journey on horseback, for there were ruins on the way. The main party returned to Red Lake and switchbacked to the Pass. The roads had been damaged by late rains and both parties were a day late.

The trader at Red Lake opened house for us and between the stores of the trader and our camp chest it was something like a return to civilization. The living room above the store was well equipped and the ladies took possession, the gentlemen making their nests in the sand-dunes.

These trading posts are constructed much on the plan of the old frontier forts. The buildings are strong, the counters high and sometimes screened, for in their trade discussions the Navajos may resort to direct action. A few traders have lost their lives in these disputes and some of their goods. One of these was an elder brother of Wetherill. We look back with much pleasure to the over night at Red Lake.

The road to Marsh Pass led through the Klethia valley. Lake reservoirs, fields and corrals by the road side, luxuriant sunflowers and fire-weeds promising greater agricultural development, is our recollection of the ride. Marsh Pass is a rocky cut between the Black and Skeleton Mesas. An abundance of firewood and water stored in natural cisterns make this a convenient camping place, and Leslie kept camp while the entire class on foot explored the ruins for a couple of days in Laguna Canyon.

It was a pleasing journey of six miles along the floor of the canyon with high cliffs and palisades to the noted Betatakin ruins of 148 rooms. A rain storm overtook the lagging snail party, and while they were crouching under overhanging cliffs they were given an exhibition of many bridal-veil falls breaking over the precipices. The forest dooryard at Betatakin was somewhat damp the remainder of the day, but the quaking asps and spruce were swarming with Pupillidæ, and here was found something new, *Pupilla hebes* mut. *albescens*. The damp collectors by a fire and protected by the city arch slept the sleep of the honest toiler and dried their clothing. The ladies descended ladders from the roofs and spread their blankets on the smooth sandstone flooring. The gentlemen slept on rocky shelving above the houses and the Dean, Casabianca to the core, stayed by the cooking beans and got wet.

"Betatakin is a homelike spot," is the first thought of the visitor. The arched cavern in the cliff is 400 feet in width,

460 in heighth, opening to the west, has an easy approach, a spring of excellent water at the base, a heavy forest and a wall five hundred feet or more high, and a small stream of water in front. It seems the most delightful and romantic of situations for village life. The ruins have been partially restored by the government, and our class for the coming summer propose to make it their home while exploring a number of newly-discovered ruins near by. Supplies will be assembled at Kayenta.

A return to the main branch of Laguna Canyon and a walk of eight or ten miles further from camp the following day in which the party was somewhat delayed and strung out by the ripe currants along the trail, brought us to the Keet-Seel ruins. This city has about the same number of rooms as Betatakin, the arch was about the same, but faced east. The forest was not as heavy, the water not as convenient, it had not been restored as it should be, and access was a little difficult. The approach is negotiated by steps cut in a deep slope of sandstone for about forty feet with a hand-rail laid flat on the surface for safety. Thus those who approach must come humbly on all fours. The pottery was a rich find at these ruins and there still remain many wagon loads of the broken material.

The probabilities are that the Hopis were compelled to leave these delightful homes against their will; that they were too easily penned up here by the war-like Navajos and their Apache cousins. At least the Hopis now live on the small and high mesas of the desert where they can see out in every direction, watch their flocks and fields and get a fair view of all who approach. Such is the theory. The decorations upon pottery, the architecture of dwellings and community buildings, with timber, corn and pumpkin rinds preserved these hundreds of years by the overhanging arches, are substantially the same as those now in use by the Hopis of Moenkopi and Walpi.

Upon the return journey bones of *Lymnæa* in the bed of the creek, in banks, washes and ant-hills above started an investigation, and it was found that these shells were imbedded in a streak of marl and peat soil sometimes a dozen feet below the canyon floor. Wetherill told us that thirty-five years ago the valley contained a chain of swamps fed by the stream. A similar condition was found at Fredonia, Arizona, by Ferriss and Daniels in 1910. The older residents said that twenty-five years before the Kanab Wash was clothed with grass and there was merely a few damp spots here and there along the valley, that the cattle had cut a trail down the valley and this trail had been deepened year after year by the stream. In 1910 the water of Kanab Wash was 90 feet below the floor of the valley and a permanent stream was of such a volume as to be known as a river. A recent freshet had taken out the community dams storing water for domestic use at Kanab and Fredonia, and the village streets were still muddy from the disaster. Perhaps these two streams and many others had a big cut the same season and by the same freshet. We see much evidence of this cutting and also of some filling. Perhaps after a stream here is cut to the bed rock it again fills with brush wood and soil washed from above.

Lymnæa stagnalis appressa Say was found in the canyon peat and it may perhaps still be found alive in some of the ponds and lakes of the mesas. We saw the lakes but an auto party is too fast for pond snails. *Lymnæa (Galba) palustris* (Müll). *Lymnæa proxima* Lea, *Lymnæa (Pseudogalba) parva, Planorbis trivolvis* Say, and *Succinea retusa* Lea, now a stranger to the locality, were also gathered; but the material was in poor condition, the shades of night were coming fast, all alone in an Indian country and it had been a twelve-hour walk. It was ten before the camp fire was beckoning at Marsh Pass.

Wetherill came to escort us to his home at Kayenta the next morning, and then led us two more days on horseback through Monument Park where peaks, steeples and effigies more than a thousand feet high seem to stick up through the plateau floor. While waiting for the snake dance, nearly a week of delight in desert literature, paintings, photographs and evening lectures was our lot at this club-like home. It is something of a headquarters for the government explorers, and for all sorts of writers, artists and students who desire to know something of the Navajos.

The party divided here, one half returning home, the other going on to the snake dance at Walpi. The journey of two days by auto was made in five, owing to weather conditions. We enjoyed the journey through the Chinlee valley where with government assistance thousands of acres of corn were under cultivation, and the side-winder rattler was added to our collection.

We also stumbled into Ganado, headquarters of the Hubbel string of trading posts established some forty years ago. Hon. Lorenzo Hubbell, its head, many years a representative of the territories of New Mexico and Arizona in Congress, was at home. Here was another museum of Indian baskets, blankets, paintings, desert books and the many things Indian we were looking for. Paintings of all the patterns in blankets used by the Navajos were on the walls, and one hundred at least of the original portraits in sepia of Indians by that best of artists, Elbridge Ayer Burbank.

Lorenzo Hubbell, Jr., of Oraibi, was a delightful acquaintance. In an empty Buick he overtook us the next morning after the Ganado visit. "Throw in a lot of those dunnage bags and some of those girls and I will help you the next ten miles; the road is rough that far," he said; and we went to it and built a bridge. When the flood from the cloudburst had passed we ran ahead into another cloudburst and built another bridge, the men folks, including Hubbell, pulled off their shoes, rolled up their pantaloons and waded through the mud and cactus for half a day in their bare feet, built bridges, dug out machines with shovels and their bare hands, pushed and slipped and tumbled until dark, and Hubbell stayed with us through it all. He was plainly that kind. When the cowboys and Indians saw him at a distance they grinned the width of their face, came up, slipped off their horses and shook hands heartily.

Humiliating to relate, an Indian boy with a burro was employed to pull out a car we could not push, and did it. On another occasion two men of our party, stuck upon the hillside of the San Juan, had their machine pulled over the top by a Najavo woman and her burro, with merely a rope around the donkey's neck.

The snake dance of the Hopis terminates an annual nine-day

religious ceremony, a prayer for rain. Here were seven hundred spectators from coast to coast, as interested and respectful as these deeply religious Indians themselves. About sixty or seventy live snakes were carried around the ring in the mouths of the priests, one snake at a time. Twenty or more of these exhibits were the common poisonous rattler—the side-winder or Edwards' massasauga (*Sistruris catenatus edwardsi* B. & G.), and the other the prairie rattler (*Crotalus confluentus* Say). No fangs were pulled, no persons bitten, no fainting, none were awe-stricken. There was no frenzy. Everybody cool and satisfied. Even those who paid a dollar for a watermelon or fifty cents for a loaf of bread ate calmly, politely and said nothing.

The party again divided at Holbrook, and at Galup Mr. Clute left for home and Cummings and Ferriss made a side trip to Montecello, Utah, via. the Ship Rock agency and Cortez, Colorado, thus avoiding the Ute Mountain and passing over the toes of Mesa Verde with its great ruins.

The Blue Range, known on some of the maps as the Altas Abajo, is about eight miles from Montecello. The walking is good and the lumber road lands one at the sawmill on the north fork of the Montezuma Creek, the very heart of the mountain range. These peaks are covered by thick groves of aspen and spruce with large open spaces of coarse grass and slides of sandstone fringed with wild currants and raspberries. Here again *Oreohelix y. cummingsi* was found abundant in the shale and also scattered among the rock slides and the aspens, with *O. cooperi* and *O. depressa*. At station 365 a few *cummingsi* were found approaching the albino form. At station 370 in tall grass *O. cooperi* was variable in size, also in the same environment in the vicinity of the copper mines, our Sta. 366. As a rule these were much smaller than those found in the aspens.

September 13th the party again divided, Ferriss to Joliet and the Dean for Tucson, taking with him a couple of young Wetherills to the University, adding with his machine that much to our desert journey. The girls did their part like men, there was no sickness, no accidents, no great adventures and it was the most enjoyable picnic ever in the most country per acre ever.

A SEARCH FOR LIGUUS.

BY CHARLES TORREY SIMPSON.

No more zealous group of shell collectors exist than the lovers of the Liguus tree snails of southern Florida, Cuba and Haiti. Many colorful races and hybrids exist in the Florida hammocks, which are wooded, raised "islands" in the swampy Everglades. The first Florida race was described by Thomas Say in 1825. The first good summary was published in 1929 by Charles T. Simpson who is also author of The Nautilus *article; "A Search for Liguus." The most complete monograph of the Florida races was published in 1946 by Henry A. Pilsbry in volume two, part one, of his* Land Mollusca of North America.

To capture the extent of "Lig" fever felt by many people, I have added a series of short articles on the mystery of the blue Liguus at Marco, Florida, and Dr. William J. Clench's enthusiastic account of "Ligging in the Everglades of Florida" with Pilsbry in 1931. The National Everglades Park referred to in Clench's article was established in 1934, and today serves as a sanctuary for many races of Liguus.

Charles Torrey Simpson was an excellent writer and prodigious worker, as well as a well-rounded naturalist. He was as well-versed in botany as conchology. He was born on June 3, 1846, at Tiskilwa, Illinois, but spent most of his life in Florida. For thirteen years he was on the scientific staff of the U.S. National Museum. He produced a monumental series of tomes on the freshwater mussels of North America; co-authored with W. H. Dall the famous 1901 Mollusca of Porto Rico; *wrote extensively on tree snails; and contributed 44 articles to* The Nautilus.

Simpson received an honorary doctorate in 1927 from the University of Miami. He died Dec. 17, 1932, at the age of 86, at his home in Lemon City, (now North Miami), Florida.

For a considerable time in the past I have been making annual trips to the Florida Keys for the purpose of studying the life of the region, its geographical distribution and the geology. Sometimes I have gone by boat but oftener by train, running to the most southern point visited and tramping back. I formerly went alone carrying no load and like an invading army trusting for sustenance on the territory I visited; but I have been so regularly taken for a tramp or bad man and driven from the doors of the natives, that of late I carry a small tent, bedding, provision, even drinking water, and by that means I am independent and can camp whenever and wherever night overtakes me.

Several islands of the lower chain have a considerable growth of the Carribean pine (Pinus caribaea), found generally in the southernmost part of the State. Big Pine Key is pretty well clothed with this kind of forest; and it is found on No Name, Little Pine, Cudjoe and several other keys.

Big Pine is a sort of headquarters from which I make trips to nearby islands. It is the largest of the lower keys, being over eight miles in length and about two and a half in width at its widest part. It runs from north northwest to south southeast and in shape reminds one somewhat of one of the modern Ku Klux clothed in full regalia, its robe flowing irregularly down its body and ending above in a comical, twisted headgear. There are two projections on its eastern side, the southernmost being a considerable island at high tide but connected when the water is low, and the whole is called "Doctor's harm" (arm). A long strip of swamp stretches southward from its southeast corner which suddenly turns to the westward and the projection is called "The helbow." All the island except the long strip on the south is oolitic limestone, the latter part being an old coral reef and a part of the upper chain which ends with the nearby Newfound Harbor Keys.

One who visits the islands for the purpose of collecting tree snails may be said to almost be "between the devil and the deep sea." It is probable that not more than 35 inches of rain fall yearly on a considerable area of the Lower Keys, and the greater part of this comes in the warmer season. During the drier part of the year what few arboreal snails still remain on the keys hide away in holes or crevices of the trees or even deep under rocks so that it is well nigh impossible to find them. This is the period which is supposed to be free from mosquitos and the only one during which a collector can have any comfort. During the warm season when most of the rain falls the keys are generally an inferno caused by these insects. Sand flies are in order at all times of the year. I generally go the latter part of October, hoping to find the mosquitos departing and the snails still somewhat active, and come back early in November.

On a recent visit to the keys I stopped as usual on Big Pine and made my way back on the railroad to the "helbow" where a noble piece of hammock, covering perhaps forty acres or more, once stood. Part of it was long ago cleared and planted but later abandoned. Charcoal burners have cut the best of the timber for their business and hurricanes have wrought great destruction in it, as it is in a badly exposed locality.

Between the railroad embankment and the hammock a tideway, about twenty feet wide and three feet deep, drained the great swamp into the open sea. I would either have to wade it and get my clothes wet or take off trowsers, shoes and leggins and on my return go through the same operation. I set my wits to work to contrive a bridge from the timbers which were thickly scattered about by a former hurricane. I laid a track of plank, got a piece of an oar and a broken gaff out of which I made a couple of rollers. Then I strained and lifted onto these a twenty-foot timber, six by twelve, which had done duty in some old railroad bridge, and rolled it down into the water. I shoved the far end of it into a little cove on the opposite side, staked the near end to keep it from drifting away and triumphantly walked across it, saying to myself, "When a man uses his brains he can save himself a lot of discomfort." In front of me grew perhaps a half acre of saltwort (Batis maritima), a dense, half-erect shrub with very succulent leaves, and I strode through this on my way to the sandy shore beyond. Suddenly I bogged down, going over my knees into water and mud that the deceptive shrub had entirely concealed, and after floundering across a couple of rods of this loblolly I crawled out on the opposite side completely bedraggled and disgusted. I reached the sandy shore and a little farther on the hammock. This piece of forest is doubtless classic ground. In the first half of the nineteenth century there lived in Key West a Dr. John Blodgett, who practiced medicine and carried on a drug store. He became greatly interested in the botany of the keys and made collecting trips among them. He discovered two Clusias, tropical strangling trees, and a Cupania, a member of the soapberry family on Big Pine, and as this hammock was very accessible to any one coming from Key West he no doubt collected in it and in all probability discovered these trees in it. I have searched the forests of Big Pine, and Dr. John K. Small of the New York Botanical Garden has done likewise, but no vestige of any of them has been found, and they are probably extinct so far as our flora is concerned. Henry Hemphill, perhaps the best conchological collector of his time, worked, I believe, on this key (perhaps in this hammock) and the adjoining No Name and found beautiful Liguus solidus in variety.

Without a doubt these snails have lived in this hammock until lately, perhaps until the dreadfully disastrous hurricane of September, 1919. I visited it a couple of months later and

A New Catalogue of North American Land Shells.

A revised reprint of the Catalogue published in THE NAUTILUS from August, 1897, to April, 1898. Giving the geographic distribution, most recent synonymy, varieties, and classification of all species known to inhabit America north of Mexico.

35 pages, in paper cover. PRICE, 10 Cts.

Address, C. W. JOHNSON,
WAGNER FREE INSTITUTE, PHILADELPHIA, PA.

found many dead and broken Liguus along the shore in front of it, some of them still well colored, probably washed out by the exceedingly high tide which covered much of the floor of the forest with sand and debris. At the time of my last visit I spent the better part of a day carefully combing it over in the hope of finding this snail alive, but in vain. My search only brought to light a few dead, faded shells inhabited by hermit crabs.

This hammock although nearly ruined seems to be headquarters for cacti on the Lower Keys. Chapman's "Flora of the Southeastern States" only gives six species, two of which are introduced, for the entire region covered by his book, but Dr. Small lists no less than eight natives from this hammock alone. Among them is a tall, columnar Cereus with trunks as large as a man's body and twenty feet high, and another more slender but erect form which I discovered on Lower Matecumbe Key several years ago. One of the prickly pears (Opuntia) is nearly prostate and has joints about the size and shape of an old-fashioned hunting-case watch—a most striking form. In fact whatever time and attention one is not compelled, when in this hammock, to give to fighting mosquitos and sand flies, must be devoted to crawling through and avoiding cacti.

A few days later I was joined by my friend Dr. Edward Mercer, formerly of Philadelphia but now of Miami, who has been with me on several recent trips. In the village of Big Pine I was told of a man, who, not long before, had found *Liguus solidus* in variety on the northeastern part of the island where he had gathered a quantity and could have taken a "hatful." That has become a stereotyped word, and every time I visit the Lower Keys I am told of some one who could have filled his hat with them. And when traced down it turns out that he has perhaps gotten a few Oxystyla, which still sparingly persists on some of these islands, or *Drymaeus multilineatus*, a very abundant but much smaller form. In some cases the bona fide sworn-to Liguus turns out to be *Litorina angulifera*. I have come to believe that the spot where *Liguus solidus* is abundant is either at the end of the rainbow or where you pick up the will-o-the-wisp. One man volunteered for a consideration to guide us to the exact hammock where these snails had been found, but when we had arrived at a couple of tumbledown houses in the northern part of the island he didn't know just where it was, but swept his hand around the horizon in a vague way and said the hammock was "off yonder." We found nothing, not even a bone.

We determined to tramp across the island and make an attempt to find Watson's Hammock which lay on the opposite shore. We had been told that we would find the walking fairly good and were given the general direction. In two minutes we ran into a buttonwood swamp which I have since learned covers the greater part of the interior of the island. This is not the buttonwood or sycamore of the northern states, or any kin to it, but a tropical tree with dark, greenish, very combustible wood which inhabits brackish swamps or their immediate vicinity. It is a strange tree, having many forms, sometimes erect with a height of 70 feet and a trunk diameter of two feet; again it falls over and becomes a gigantic, writhing half-vine. On drier ground it is a small, somewhat erect tree, and in this swamp it grew in this fashion, only it threw out a good many stiff, crooked branches just at the ground which admirably served to trip our tired feet. It was only a short time until we came to more or less extensive pools and ponds of beer-colored, brackish water, which we tried for awhile to avoid by making a circuitous tramp around them. Soon, however, it became apparent that we must wade, and we plunged in, often to the depth of three feet, blundering and even falling over the very irregular bottom. Then for a long distance we encountered stands of buttonwood, dense scrub hammock and water. This hammock was, without exception, the most difficult to get through I ever saw. A considerable part of it consisted of a small tree or large shrub, a Bumelia or ant's wood, with narrow leaves and innumerable branches. The whole purpose of the tree seemed to be to develop and carry an immense load of long, excessively sharp thorns which for their ability to catch hold and hang on cannot be surpassed anywhere. It formed thickets, not quite as dense as a haystack, but the next thing to it, and we could no more crawl through it than we could through the side of a battleship.

This growth which belongs in slightly brackish ground bordered the hammocks and we had to get through it in some way to get across the latter. Often we got into a pocket and after fighting our way along for a while we were obliged to turn back and get out the way we came in. At other times I got down and cut my way through with my pocket knife so that we could push our bags ahead and crawl after. In the real dry hammocks the Bumelia was replaced by the pull-and-haul-back (Pisonia aculeata), and a tropical prickly ash (Zanthoxylum), to such an extent that they were nearly impassable. Wherever we found hammock I strained my eyes to find Liguus but saw none.

There was only the ordinary development of sand flies and common gray mosquitos, but we had scarcely gotten into the swamp before we began to encounter swarms, or herds, or droves, whichever they might best be called, of an enormous black mosquito, the largest and most terrifying I have ever seen. They shone as if freshly varnished and came on with a steady, leisurely flight as if they were sure of their victims. The fore part of these monsters bent down in a remarkable way, probably to allow the proboscis to get into action for some time before the rest of the insect arrived. The doctor at once called them "Dirigibles" which we soon shortened to "Blimps" on account of the inconvenience of using a word of four syllables whenever we encountered them. When one of these became filled with blood and slowly sailed away with its various appendages trailing below and after, it suggested a zeppelin in a remarkable manner.

For five dreadful hours we fought our way through this inferno. Often the growth was so dense that we could not see the sun and we constantly consulted our pocket compasses and bore off to the west or northwest whenever it was possible. Sometime before sundown I saw an open spot in front, then I caught a glimpse of the sea and a date palm which some one had long ago planted near the shore. In a moment we stepped out onto a level, smooth, grass-covered prairie that stretched to the Torch Key Channel, and we swung our hats and capered about like boys. Taken all in all, I believe this was about the most difficult short tramp I ever made, and when we got near the shore I was glad to throw myself down on the grass and rest.

The doctor is a delicate man of 60 whose health is none the best, and when he first proposed to go tramping with me I felt very doubtful whether he would be able to endure the hardships of such a rough-and-tumble life. Instead of lying down and resting that evening he took a long walk through the pine woods apparently for mere relaxation. In our various excursions I found him always ready to lead, and he never gave any intimation that he felt the slightest fatigue.

We pitched our tent on a growth of sedges (Eleocharis), so dense and tall that we could walk on it without pressing it to the ground—an admirable bed. Then we crawled in and after driving out the mosquitos carefully closed, as we thought, every aperture and congratulated ourselves on the prospect of a fine sleep. I did sleep but uneasily and was vaguely conscious that the doctor was much disturbed. When daylight came we could see that the inside of the walls and roof of the tent were so covered with mosquitos gorged with our blood that in places one could not have put his finger on them without touching an insect. We discovered a minute opening, just large enough to admit them in single file, and they had been industriously passing in all night.

We searched some small hammocks which lay to the northwest of us without results and then turned southward along the shore, finally reaching the great Watson's Hammock. Formerly this covered a considerable area and consisted of a magnificent growth of tall, closely-set tropical trees. Much of it has been cut out; but there still remains a splendid remnant, and this we diligently searched but found only a few faded bones. I first visited this forest in 1885, arriving just before sundown. My boatman who was anxious to get on only consented to allowing me a few minutes on shore, but during that brief time I could see that the trees were full of splendid Liguus. Had I been allowed a little time I could have actually gotten "a hatful." As it was I found the type of my *Liguus solidus crassus*, a form with a very solid shell and truncate columella, ivory white with a narrow bronze-green peripheral line, also a couple of specimens of the form *graphicus*.

We struck off down the island in a general southeast direction, but in trying to avoid the swamps were obliged to zigzag about considerably. The walking on the lower islands is much better than that of the upper ones, the general surface being level and comparatively smooth. This is an oolitic limestone much like that of the Miami region, but it was deposited in a shallow sea while the latter formed a retreating shore and is irregularly stratified. In many places the rock of the Lower Keys has split loose in thin layers and become broken up, and between the pine trees there is often a dense growth of a palmetto (Thrinax microcarpa). The whole is generally tied together with a villainous climbing smilax which is most liberally provided with thorns.

After a long tramp we reached the village, and the next morning had my old friend Joseph Sears take us across the strait to No Name Key. Sears is a powerful Bahama negro, good-natured and voluble, and in time past he has taken me to many of the keys in his boat, the "Three Fannies."

We camped in the front yard of an abandoned place, again getting a bed of the Eleocharis, and this time were fortunate enough to entirely shut out the mosquitos, but we could hear their angry humming all night, music which lulled us to sleep. No Name is nearly three miles long and about a mile wide, its northern part being pine forest with a dense undergrowth of palmettos. There is, or has been, a great central hammock of magnificent tall, closely-set tropical trees, but about 60 acres of it have been cut out. Much of the southern part of the island has been hammock but most of it is now destroyed. We carefully searched this interior forest, almost tree by tree, but found no living Liguus. In places the ground was thickly strewn with broken and bleached shells, some of the fragments still retaining color.

It may be asked, "What has become of them?" Their disappearance is due to several causes, man being the chief. The building of the extension of the Florida East Coast Railway wrought terrible destruction among the hammocks on the keys. One of the finest pieces of this growth was located on Key Largo where sparks from the engines set the timber cleared from the right-of-way on fire and destroyed hundred of acres of splendid

Liguus-bearing forest. There has been a series of years with a deficiency of rainfall, twelve in number according to a well-informed settler on one of the keys, and the snails have been driven to take refuge in crevices of trees and under the rocks while doubtless many have been exterminated. Birds have killed many more. Possibly dry weather has made food scarce and caused the birds to prey to a greater extent on the snails than usual.

I am strongly inclined to believe that a few colonies of the *solidus* may still exist on the Lower Keys. Three years ago I found three adult specimens of the typical form of this on trees back of the village of Big Pine. Several years ago Henderson and Clapp found a large colony of *solidulus* on Stock Island, but at my last visit to this only a few dead shells could be obtained. Within a couple of years I have found several tolerably fresh shells of *graphicus* on No Name, two on Sugarloaf and a fairly good young *solidulus* on Summerland and Boca Chica. There is still a good deal of hammock, some of it second growth according to my friend, J. T. Knowles, of Big Pine Key, on several of the islands, and the fact that settlement is decreasing rather than extending on the lower chain makes it probable that this growth may spread. If we could have a series of wet years it is easily possible that these snails might increase and be abundant again in places. There is no group of land snails on earth with more wonderfully beautiful shells than those of *Liguus solidus*. Their texture is of a marvellously delicate porcelain, their polish is remarkable and the colors of some are bizarre and extremely rich.

NOTE.—The above forms part of a chapter of my forthcoming book entitled "Out of Doors in Florida."

JUST OUT.—CONTRIBUTIONS TO THE MOLLUSCA OF FLORIDA.
By CHAS. T. SIMPSON.   Price 25c., postpaid.
Nomenclature and Check-List of American Land Shells.
By H. A. Pilsbry.   Price, 10c., postpaid.   50c. per dozen copies, postpaid.
Address W. D. AVERELL, - - - - - - - Mt. Airy, Phila

"LIGGING" IN THE EVERGLADES OF FLORIDA

WILLIAM J. CLENCH
Museum of Comparative Zoology

With very few exceptions, the collector of arboreal mollusks has more thrills crowded into one hour of good hunting than a general collecting trip will yield him in a week. To locate a hammock containing a colony of *Liguus* is to reach a high spot in collecting that has but few equals.

Such was our good fortune during the past winter. I prevailed upon W. E. Schevill, a fossil hunter, to drop his antiquities long enough to enjoy a bit of modern zoology in the land of sunshine and mosquitoes. The Museum truck was at our disposal and equipped with our camping duffle we set out from Cambridge headed for Florida on Feb. 9th, in a delightful New England sleet and snow storm.

Noon of the following day we stopped at the Academy of Natural Sciences in Philadelphia to see Dr. Pilsbry. Since the Pinchot trip, Dr. Pilsbry has scarcely been on speaking terms with me. After all, a seven months' expedition to the South Seas is pretty tony, and my ravings of the Tennessee and Georgia trips I had previously made left him rather cold. Fortune was with us. The day was cloudy and the worst part of winter was still ahead—and that's a lot of grief for anyone living in Philadelphia. We rubbed it in. The joy of collecting in new country, warm weather and old clothes, have more than ordinary appeal for the Chief, and we handed it to him. At the end of an hour's talk we could detect real envy, and considering the books partially balanced we left for Washington and beyond.

EXCHANGES.

Litiopa striata from Sargasso Sea; Donax Cayennensis; denticulata; Macrocerramus Klatteanus; lineatus var. (n. s.) Hayti, fine; Unio Canadensis, etc., etc., all my own collecting. Lists exchanged. J. J. BROWN, M. D., Sheboygan, Wis.

Offered:—Helix nemoralis and Helicina occulta from Rockbridge Co., Va., for shells from other localities. JAS. H. MORRISON, Box 34, Lexington, Va.

Four days of good roads, warmer weather and car trouble saw us entrenched at the home of Dr. Fairchild in Coconut Grove. Here we were able to outfit more completely for our camping trip and enjoy a few days of delightful hospitality on Biscayne Bay before entering the Everglades.

Our original plans had called for a boat trip through the 'glades from Pinecrest south to Long Pine Key. However, the water was too low for such transportation and we decided to go to Long Pine Key, which was but imperfectly known. This is an area a little west of Royal Palm State Park. It is composed of glades alternating with pine ridges or islands. A new road eight miles long has been constructed running west across about half the length of the key, and we were to work the territory along this road. A few tomato growers farming the glades are engaged in a rather precarious livelihood, and from what we saw and learned, the plowing must be done with a motor boat! When we arrived the country was fairly dry, though several hammocks, especially south of the road, had to be reached by wading in water six inches to a foot in depth.

Through the courtesy of Mr. Frank van Marlen, the manager of the park, permission was granted us to camp near the park lodge. From here we could drive out each morning to our collecting grounds leaving our camp set up.

As stated above, only a few hammocks in this region had been investigated. Most of these were near the road and readily accessible by walking up the glades. Our introduction to these hammocks proved them to be far richer than we had hoped and altogether too good to be enjoyed alone, so letters were sent to Dr. Pilsbry and Mr. Goodrich to join the party. Mr. Goodrich found it impossible to come, but Dr. Pilsbry telegraphed that he would be with us as soon as the southbound train could carry him. He never told me what excuse he gave the powers that be in Philadelphia for going to Florida in mid-winter. I find it exceedingly difficult to explain why the best "lig" collecting is to be found during the height of the gay season!! Tsk! Tsk!

By the time Dr. Pilsbry had arrived we had amassed a

sizable fortune in *Liguus,* and during the auto drive from Homestead to camp, vivid stories of collecting were detailed at length.

We had made arrangements for Dr. Pilsbry to sleep at the lodge but to have his meals with us, as our particular type of tent could only accommodate two people. This arrangement worked splendidly for all hands.

Hammocks, as these thickets of broad-leaved trees are called, were found everywhere: in the pine woods, the open glades or on the edge of both the pine land and glades. In general they were quite alike other than in size. The flora of the hammocks is essentially West Indian, with the *Ficus, Lysiloma, Bursera* abundant and with them hosts of other plants unknown to us. The best trees for *Liguus* in this region were the *Lysilomas,* and wherever these trees were found the "ligs" were sure to be found as well. At this season the leaves of the Lysilomas are dark red and this color could be seen for some distance. This aided materially in locating "lig" hammocks, as their colors ordinarily blend with the greens of the pine woods. The story of each day is much the same. Each daily trip had, of course, its own delight in locating different color forms or colonies of *Liguus* that differed in some way from all others that we had seen.

One Sunday Richard Deckert and Henry Frampton of Miami came down and we all started out on our longest single tramp, penetrating the 'glades six miles north of the road. This trip yielded three pure races of *Liguus,* namely *L. fasciatus luteus* Simp., *L. f. castaneozonatus* Pils. and *L. f. cingulatus* Simp., each in a single isolated hammock in the everglade. It is the occurrence of these pure colonies that show the fixity of these color forms as Mendelian segregants.

On two other occasions Paul McGinty and his sons Paul and Tom from Boynton spent the day collecting with us. Maxwell Smith was with the McGintys on their first trip and that day we drove to the end of the road and then tramped due west for nearly two miles to visit two large hammocks. The first contained several large royal palms but the hammock growth in general was so dense that *Liguus* were rather difficult to find. Charles Mosier made several trips with us and directed us as well to many hammocks that possessed some fine material. Dr. and Mrs. DeBoe of Miami, friends of Dr. Pilsbry and also ardent "lig hunters", drove down for collecting two or three times during our stay. Their trips were made independent of ours but we had the pleasure of comparing notes and exhibiting material that had been collected during the day.

Through a photographer in Miami we had made a series of aerial photographs of our section of the Key. The day the photos were made was a good "flying day" though not so good for photography. Sufficient haze existed to limit materially the height at which pictures could be made and consequently the pictures did not exhibit enough detail to locate many hammocks more than a mile or so from the road. The road had to be included in the pictures to serve as a guide, as this flat country offers but little by way of landmarks. The pictures covered, roughly, a strip about ten miles long, two miles beyond the limit of the eight-mile road. We were able to scale the pictures to distances on the road and readily locate any hammocks that were noted in the photos.

Hammocks could be located blindly by just walking through the pine woods, but this is slow and tiresome work as walking is exceedingly difficult over the rough limestone rock which is pitted with holes and always badly weathered.

Fortunately for us, rattlesnakes were rare in this region. Only four were encountered, three of which we managed to capture, but not alive! It is quite disconcerting to have your eyes focused on a *Liguus* a few feet away and then be suddenly alarmed by the "buzzing" of a rattler. According to popular belief, a rattler always rattles when disturbed, but *Crotalus* No. 3 failed to do so even when being walked over, around and almost upon—if two inches can be called close enough to be considered a disturbance.

Mosquitoes were very rare until the last week and then they became quite abundant, enough so that supper usually consisted in part of setting up exercises. Also a very large Tabanid fly appeared during the last week. It's buzz could be heard fifty feet away. Visions of the northern "green head" came to mind, and if the bite of these fellows was at all in proportion to their size it must be equal to that of a dog. Fortunately their approach was heralded by so much noise that full preparations could be made to drive them away or else kill them after they landed and before they had managed to find suitable territory to puncture.

It is unfortunate to record that this area as well as others in the Everglade region is being rapidly destroyed. Drainage and fire are fast despoiling a unique territory. Unless steps are soon taken to preserve it, little will remain of the original fauna and flora.

Attempts are now being made to make this section of Florida into a National Park. If this project succeeds a remarkable area will be held for the benefit of future generations, an area that has no parallel for its kind anywhere else in the world. As a wild life preserve, especially for birds, it probably has no equal in the United States, as its interior is so inaccessable, without roads, that natural conditions would prevail indefinitely.

There is an appeal for the naturalist in this country that cannot readily be described. I suppose that it is so totally different from most country that many of us have seen or lived in that it intrigues our interest. For all of its flatness, it is a region of strong contrasts. The open glades or prairies, the pine woods and the hammocks are sharply defined. A plant ecologist could find problems for a life's work in any square mile on the Key. The evolution of a hammock can be seen from a unit measure of a two-foot willow tree in a small sink hole to that of Paradise hammock, a square mile in area with several hundred species of plants. To the malacologist, it leaves little to be desired. Problems of speciation in the color forms of *Liguus*, methods of dispersal, adaptation, local distribution and life histories offer a multitude of studies.

We had noted numerous empty shells of *Ampullaria paludosa* Say on the canal banks near camp. No live snails had been seen in the canal though a few had been found in the solution or sink holes in the glades. One night rather late, after a feast of grapefruit we proceeded to the canal with the flashlight to wash up. A pair of live *Ampullaria* were found slowly crawling along the bank. Investigation of two hundred feet of canal yielded over 100 specimens. During the day this mollusk had been buried in the mud and algae in the deep water of the canal. At night, protected by darkness, it could feed at leisure along the bank. The presence of many large birds that feed on them has probably been a factor in producing this habit. Elsewhere in Florida these snails have been easily found during the daytime.

Ampullaria must be an important food item for many birds if their dead shells can be considered an index. Lang[1] has published an account of the feeding habits of *Rostrhamus sociabilis*, the Everglades Kite on *Ampullaria* in British Guiana and in all probability the same thing occurs with the kites in the Everglades. *Planorbis intercalaris* Pils. must as well offer an abundant food supply for birds as it is very common in many sections of the open glades.

After five successful weeks we drove to Miami, Schevill and I returning directly by boat to New York. Dr. Pilsbry remained over a few days with relatives in Coconut Grove.

There is much country still unknown and we have hopes that in the future more trips can be made for *Liguus* exploration. Frankly, "ligitis" is a serious malady and when once infected a complete cure is impossible. It is seasonal and has a definite periodicity concurrent with the drying up of the glades, the disappearance of the mosquitoes and the first heavy snow storm in the north.

---

[1] Lang, H. NAUTILUS 37; pp. 73-77, 1924.

THE CHARLES TORREY SIMPSON PARK.—Those who have hunted *Liguus* in the old time Brickell Hammock at Miami, before the marvellous expansion of that city over it, will be glad to hear that a part of the old hammock which remains unchanged has been made a city park, dedicated on April first to Dr. Simpson. It is a well-deserved honor to a veteran naturalist.—H. A. P.

Fording for Freshwater Snails

Although collecting freshwater snails was a very popular pursuit during the 1800's, especially by conchologists like Jared P. Kirtland (1793-1877), Annie E. Law (1842-1889), and James Lewis (1850-1890), there are few good accounts of fluviatile mollusks in The Nautilus *until after World War I. As pollution continued its inexorable destruction of the ecology of the lakes and rivers, interest in freshwater collecting began anew, mainly sparked by men like Bryant Walker (1856-1936) of Detroit, Calvin Goodrich (1874-1954) of Ann Arbor, Michigan, and Prof. William J. Clench (1897-) who was, at the time of the writing of the two articles included here, custodian of the Grand Rapids Public Museum.*

VAGABONDING FOR SHELLS

BY P. S. REMINGTON, JR. AND W. J. CLENCH

P. Sheldon Remington, Jr. (1899-1975) was a boyhood friend of Clench's and joined the 1924 "Vagabonding for Shells" at the age of 25. Remington was an ardent amateur conchologist who contributed several articles to The Nautilus. *He was a teacher and a principal of several private schools in Virginia, Missouri and Greenwich, Conn. In the chapter of "Collecting Abroad", Remington's articles on "Rambles of a Midshipman" are reproduced.*

The authors have been impelled to write this article because they feel that they have accomplished something of value to conchology, and because they hope that their account of it may induce others to attempt similar trips. The "expedition" was made under the auspices of the Museum of Zoology at the University of Michigan, the expenses of the trip being borne by the Museum and by Dr. Bryant Walker of Detroit. The greater part of the material collected will eventually go to the collections of the above named. Thanks are due to Dr. Walker and to Mr. Calvin Goodrich for help in the way of advice, suggestions as to the best places to collect, and gifts of maps and other equipment. A formal report of the species collected will come out later.

The purposes of the trip were several. In the first place, we wanted to collect as large a series of Physa as possible to provide more material for Clench's doctoral thesis, which is on the Physa of the U. S. east of the Rocky Mts. We wished to get our Physa in alcohol for anatomical study, for there is a scarcity of such material. Secondly, we were anxious to collect through the Tennessee River system, particularly the upper portion, which neither of us had ever seen, and which is one of the richest collecting fields in the country. We also had a list of various places, furnished us by Mr. Goodrich, which was desirable to collect in either because they were new territory, or because they were the locality of certain doubtful species. Most of these we cleaned up. Last, but not least, we were out to get some Ios. The authors collected shells together as boys, and at that time one of the things we promised ourselves to do some time was to go on a hunt for Ios. This trip promised an opportunity of doing that and of renewing our boyhood memories. Before leaving, Clench prophesied that this year would see the last of the Ios! It wasn't quite so bad as that, but we did succeed in taking several hundred. There are undoubtedly a few left in the French Broad River, which we did not touch. We had carte-blanche as to where we might go, but were guided largely by the above considerations. We left Ann Arbor on July 19 and returned Sept. 8.

In our choice of equipment we were able to profit by the results of a similar trip made the preceding summer by Clench. We bought a 1922 Ford touring car in excellent condition and proceeded to load up the running boards and back seat until it looked like a gypsy outfit. We were handicapped by the fact that we had to carry full camping equipment and supplies as well as the collecting kit. We had a very handy auto tent, two folding cots, two folding chairs, and a folding table. This last was bought because of Clench's experience the summer before, and it was certainly useful. In our grub box we carried cooking utensils and canned food, mainly beans. Both the authors were originally from Boston. We had another large case containing wading boots, sieves, jars, boxes, vials and bags, of which one can't have too many on such a trip. A duffle bag for each of us, a suit case of "civilian" clothes, a dispatch case for papers, a case for toilet articles, a lantern, blanket rolls, water jug, emergency can of gasolene and one of oil, and a container of alcohol filled up the back. We had a very fine two-mantle gasolene lamp which was invaluable. Although the museum appropriation for this trip came largely from the conchology department, we received contributions from the insect and reptile departments, so had to carry along equipment for collecting such specimens. The insect equipment was limited to nets and cyanide jars, but we had to take a large milk can partly filled with formaldehyde to preserve reptiles. The reptiles were put in this to harden and preserve them, then injected and wrapped in gauze and shipped back.

With all this mass of equipment aboard you would say we must have spent half the day packing and unpacking. It is true that it did take us an hour or more for this operation at first, but before long we evolved a system, so that by the end of the trip we could pack in less than fifteen minutes and make camp in less time than that. We were surely a spectacle and got many a kick out of the reactions of the beholders. Vagabonds indeed!

One of the most enjoyable incidents of the whole trip came at the very start. We left Ann Arbor about nine in the morning "heading south." We collected a little, mainly *Physa*, and reached Toledo in the afternoon, where we hunted up Mr. Calvin Goodrich. He very kindly insisted on our putting up with him, so we had an opportunity to inspect his large collection of *Pleuroceridae* and "talk shop." As the catch where we were going would be largely of that family, we wanted all the "dope" we could get. We had a very enjoyable visit, all too short, for, as Mr. Goodrich remarked, it isn't often that three shell collectors get together.

The next day, Sunday, we set out bright and early bound southwest for Bluffton, Ind. We had heard unfavorable news from Mr. Goodrich that all the rivers in the south were swollen from heavy rains, and that unless they went down we would see very little good collecting. Our original intention was to strike south through Cincinnati and Cumberland Gap to the headwaters of the Tennessee River, then work down to Huntsville, Ala. and then north to Louisville. Upon receipt of this unwelcome news, we decided to reverse the itinerary, so as to reach the heavy collecting last, in the hope that by then the rivers would be low. This turned out to be a wise move.

We camped the first night near Hicksville, Ohio, a half-mile off the main road. We can hear the ringing reply, even now, of the farmer on whose land we asked permission to camp. "You bet your life you can!" he said. We want to say right here that if the camper is neat-appearing and respectful, and leaves no rubbish or disorder of any kind behind him, he will have no trouble finding a place to camp, and we were not once refused in the whole 1700 miles we traveled. It is usually better to go a little off the main road before camping. One should be very careful about water, and we carried a bottle of Halozone tablets which we usually added to the water. Country folk are always hospitable and often brought us out cake, pie and other dainties.

We were sorry to leave the wonderful concrete roads of Ohio, but we had little to complain of in Indiana. There was very little collecting from Toledo to Bluffton. On the banks of the St. Mary's River at Fort Wayne we found some fine *Succinea avara* Say and a few other land shells. At Bluffton we hunted up Mr. E. B. Williamson, the dragon-fly expert, and a member of the museum staff. We were cordially invited to make his home our headquarters for as long as we cared to stay in Bluffton. We were glad to stay two days while we rearranged our outfit, supplied what we found lacking, and collected in the vicinity. We found quite a few *Planorbis*, *Physa*, *Polygyra*, *Succinea* along the Wabash and near by. The river itself was disappointing, because it was in high water. The collection of dragonflies which Mr. Williamson and his brother have gotten together is one of the best in the world, and we learned many pointers from observing their laboratory and field methods. Such precision, accuracy and care of detail are well worthy of copy.

After bidding a cordial forewell to Mr. Williamson and his family, we headed south again. We collected our first lot of *Goniobasis* in the Noland River, two miles west of Centerville, Ind. They were probably *G. semicarinata*, Say. After that, *Pleuroceridae* formed the greater part of our catch. We arrived at Bloomington on the night of the 25th in a pouring rain. It being too wet to make camp and cook a meal in any degree of comfort, we put up at a hotel and after supper took in the local movie and wrote up our field notes. This latter was always a part of our evening duties, with a study of maps and a plan for the next day's trip.

It may be well here to mention how we kept field notes and cleaned our catch. As soon as we finished collecting at one spot, the entire catch was put in a can or bag and a slip of paper bearing a number put with it. Under the same number this entire lot of shells was entered in a field note-book, with full particulars as to ecologic data, etc. Here is a specimen entry:

"No. 103. *Pleurocera, Campeloma, Angitrema* and *Anculosa*. Little Tennessee River, Morganton Ferry, Tenn. Along muddy banks and on rocks and shoals. 8/22/24."

That, by the way, was one of the finest lots we ever collected. Our lot numbers ran serially from 1 to 145, the last collected on Sept. 4. On our next trip we shall start with No. 146. Every so often we filled the cans of the last few days' catch with alcohol, usually doing it in the evening. Then, next morning, one of us would stay in camp, empty out the alcohol and dry the lots separately in the sun, while the other collected. At the same time full labels were written out and the shells wrapped up and sent back to the museum as soon as enough for a box had accumulated. We used denatured alcohol, which, while not as good as pure alcohol, is cheap and readily obtainable locally. Later, if it is desired to examine the animal for radula, it can be soaked out. This end of the collecting was not much fun, but very necessary. We are setting it down because some one else may want to adopt it and improve on it. In accounts of collecting trips the authors usually neglect to state how the shells were cared for and sent back, and we have often wondered what methods were followed. John B. Henderson in his "Cruise of the Thos. Berrera" is one of the few exceptions.

Our methods of collecting were very simple. When you see the shells, go in after them. At first we used to pack them up along the edge, walking out on boards and so forth. But after a while we got so that we waded right in after them, shoes, puttees, and all, and our feet and lower limbs were continually wet from one stream to the next and from one day to the next. Yet we never once took cold on the whole trip. Back in camp we used to put on dry clothes, but pull the wet ones back on in the morning. If the shells were pretty far out and were plentiful, we put on swimming trunks and old boots and went on out.

We struck no really good collecting until about ten miles south of Bloomington, Ind., and then every stream was thick with *Pleurocerids*, *Physa*, and often naiades. We took ten lots of *Pleuroceridae* between Clifty Creek, Columbus, Ind., and the Ohio River at Louisville. We stopped at Mitchell, Ind., a day to inspect the curious cave formations there. In an icy cold stream issuing from Donaldson Cave, we took a shell which resembled *Goniobasis semicarinata*, Say. In the cave itself we hoped to find blind fish, but saw only one and it got away. We did collect two small salamanders. From Mitchell south, the country became very rugged and beautiful, quite a contrast to the extreme flatness of central Indiana. The good ship "Asthma", however, never failed us, in spite of the load she was carrying. In the whole 1700 miles she never gave us any real trouble, although we did change tires seven times one day! But that was on the "boulevards" of Kentucky. It was with real regret that we sold the gallant little car in Knoxville. She should have been put to pasture like all noble steeds.

We collected a fine series of *Goniobasis indianensis* Pils. in the Blue River, taking lots at several points three or four miles apart. This is the type locality for that species. Here we also collected three specimens of what looks very much like an *Anculosa*, but will be determined later. The first stage of our trip ended at Louisville, where we got our accumulated mail, stocked up on beans again, and pulled out for Cave City, Kentucky.

For the first fifteen miles after leaving Louisville we struck concrete roads and were beginning to have a high opinion of Kentucky roads and the Dixie Highway in particular. Then we hit a detour, then more detour, worse than the last, and more of it for about sixty miles. When they repair a road in Kentucky they take the whole stretch at once. Words cannot express the meanness of those roads, and we were so busy "praising" them that we forgot to take pictures of them. We passed near Lincoln's birthplace, and now we know why he left Kentucky. Furthermore we struck no collecting for about forty miles after leaving Louisville. In a way, we are grateful for those detours, though, because in taking one we were led to one of the richest collecting spots of the whole trip. As we were pounding along in the midst of the hills, we decided to stop a while and cool the engine. Right where we stopped we noticed a sign which said "Visit the dam and eat at Glenbrook Hotel," and an arrow pointing down. "Dam" sounded like water (though it was what we had been saying for many miles), and water meant shells, so down we went, and we certainly found shells. The dam and brook below were plastered with a small black *Goniobasis* and we also took several hundred fine *Physa microstoma* Say out of the pond above. They have dammed up a big spring there for power purposes. As there were also indications of land shells, we drove down and camped beside the hotel for two days while we collected. In the spring above and below the dam and in a neighboring spring we took several quarts of shells, taking care to keep the lots separate. I have never seen shells more plentiful in one spot except in First Creek at Knoxville. We also discovered to our surprise that we were only a hundred yards above Rio on the Green River, so we hastened to get a boat and collect in the river. The Green River, and it is green, was one of the places we wanted to hit, because it has not been collected in above Mammoth Cave to any extent. We took half a dozen species of *Pleuroceridae*, including *Anculosa*, and a fine lot of naiades. Also along the sides of the glen above the spring we found many land shells, *Omphalina*, *Polygyra* of many species, and others. So the detours were not an unmixed blessing; indeed, we found that to get the good shells, one must get off the good roads. So it was with a sense of time well spent that we left Rio.

At Cave City we sent our first shipment north, two big boxes, and then we went out of our way a few miles to see Mammoth Cave. It turned out that this word should be in the plural, for we found on arriving that there were half a dozen caves or more, many said to be more splendid than Old Mammoth. Acting on advice we had received, we spent a morning exploring Great Onyx Cave and count it time well spent. Here also we collected a salamander, mole crickets, and a blind beetle. It was in one of these caves that Call discovered *Carychium stygium*, but we saw none. We also collected again in the Green River here, and then pulled out and set off for Nashville, Tenn.

There was not much collecting till we struck the Cumberland River just above Nashville, and here we took a fine lot of shells, finding *Angitrema* for the first time, big, spiny ones that looked like young *Io*. This was the only day we really suffered with heat and we finally stopped collecting to run into town and cool off at a soda fountain and read our mail. This ended the second stage of our trip.

From Nashville south to Huntsville, Ala. the collecting became very rich, every stream being loaded. It seems that the forecasts and reports of high water were not borne out, for in practically every stream we could wade across and see bottom. In fact, only once on the whole trip did rain so muddy a stream that we could not collect in it. It was an ideal summer from a conchologist's standpoint, for it rained scarcely once throughout the belt we collected in for the whole two months.

The Duck River, which we crossed at Shelbyville proved a big disappointment, for we found only one shell, a naiad. Doubtless this stream is polluted right there. But one mile south of Shelbyville, in a large tributary of Duck River, we found very good collecting. Here we took our first *Anculosa* in quantity, a nodulated species. They were in what we soon found to be their characteristic situation—on the edges or underside of flat rocks in very swift water, usually at the head of a riffle.

The Elk River, one mile south of Fayetteville furnished the best collecting we had so far found. Here from the river itself and from a small slough close by we took about three gallons of *Pleurocera* and *Anculosa*. We also began to get *Campeloma* from now on. A few days before, we took the first *Vivipara* of the trip along the marshy edge of a river one mile north of Murfreesboro. Shortly after leaving this locality we crossed the line into Alabama, and came on into Huntsville. This town has been made famous by the work of many other collectors, notably H. E. Sargent, H. H. Smith and H. E. Wheeler. The latter in two excellent papers in the NAUTILUS gave a list of the shells of Monte Sano, a high plateau overlooking the town. Before going up Monte Sano we collected in the Big Spring in the heart of the town. This spring contained a puzzling species of *Goniobasis*, puzzling because Sargent had distributed it as *G. perstriata*. Goodrich recognized them as *G. nassula* and thought the two names might be synonymous. The shells we found were *G. nassula*, but we turned up the true *G. perstriata* in Spring Creek. It happened that the day we visited the spring men were cleaning out water weeds, so we simply sat down and leisurely picked a box full of the little fellows. In the afternoon we drove up Monte Sano through one of the two toll-gates left in Alabama and camped on top. Next day we devoted to scrambling over the mountain, and found many of the shells listed by Wheeler, but not all, for it had been an exceedingly dry summer and land shells seemed to be very scarce. For the first time in our experience we took *Helicina*. This ended the third stage of our journey.

Our next objective was Chattanooga, Tenn. It was a question whether to take the road up the north bank of the Tennessee, or the south bank. We finally decided on the latter because the road was reported much better. So we turned south till we hit the river at Whitesburg Ferry Landing. Here we found a great quantity of naiades embedded in new wash of the river, all clean, in perfect condition, nacre and hinge unharmed. We took a large series of each species. A little farther on we crossed the Flint River where we spent several hours collecting large numbers of *Vivipara*, *Campeloma*, *Anculosa*, and several species of *Goniobasis* and *Pleurocera*. The river was very muddy, but normally so, no doubt, because there was no trouble getting the shells. We put up for the night at New Hope and collected in Paint Rock River next day. In the deepest parts the river was only breast-high, rock bottom with a layer of silt or gravel, so we put on trunks and covered every foot of the bottom for a half mile. We are quite sure we got every species there, and a nice lot it was. As usual there were two distinct species of *Pleurocera*, probably more; one, a squat, heavy form found on rocks, the other a longer, slender species found on mud and on logs. We recognized among the naiades such old friends as *Proptera alata* Say and *Leptodea fragilis* Raf. We crossed the Tennessee at Guntersville, turned west over Bear Mountain and then headed north. Albertsville, Ala., was the farthest south we reached. There was very little further collecting till we reached Chattanooga, and in this respect we were sorry we did not take the northern road. The only notable collecting was a good series of a fine, large *Pleurocera* from Lookout Creek about four miles north of Valley Head. On Aug. 13 we rounded Lookout Mountain and entered Chattanooga. This ended the fourth stage of our trip.

At this point Paul J. Adams of Knoxville, Tenn. joined the expedition as an additional collector and stayed with us till the end of the trip. His knowledge of the country around Knoxville and the Great Smoky Mountains was extremely useful to us and enabled us to cover much more ground that we otherwise would have. On the other hand we were able to show him wrinkles about the "shell game," which were new to him. We now began to strike more of the localities which Mr. Goodrich had suggested as desirable to visit, either because these were type localities or represented only by puzzling and inadequate material. We laid out our course especially to hit such places. The first of these was Chickamauga Creek at Lee and Gordon's Mill, just outside the military park. The reason for visiting this was to check up with material from Crawfish Springs in the park which was the type locality for *Pleurocera planicostata*. We found shells plentiful there and collected several quarts. We then drove west to reach the Conasauga River, which is the type locality for eight species of *Goniobasis* alone. We were also anxious to collect in this river because it belongs to the Gulf drainage and for that reason presents a different fauna from that of the Tennessee. The shells did indeed look different from any we had so far seen and we took large lots from two places on the Conasauga, as well as Mill Creek, Coahuila River and several other streams. We recognized among these *Goniobasis murrayensis* Lea—this is its territory. By accident, while buying postcards in Dalton, Ga., we saw a picture of "Country Club Lake", which was not on the map. Arriving there we found a fine large *Pleurocera* of the lake type. Mr. Goodrich is now working at identifying all this material.

A few miles beyond Dalton we turned north again into Tennessee and went out of our way to visit the Ocoee River dam at Parksville. This dam has backed up the river for ten miles and made a large inland lake; we were interested to see what the shell life would be and we also wanted to collect in the Ocoee, as no shells have been reported from there recently, if at all. We found out why; there *are* no shells in the river. We learned that farther up the river near Ducktown, sulfuric acid was produced as a waste in zinc smelting, and run off into the river. This has completely killed all forms of water life, not even a fish being found. This affects even the Hiwassee River, into which the Ocoee flows, so that we found no shells in either. It was a somewhat new experience, for we had grown so used to finding shells in every creek that we had acquired the habit of piling out of the car, grabbing an empty can and wading right out. At times it almost got monotonous. We climbed a small mountain, Sugar Loaf Hill, beside the dam and found quite a few of the smaller land shells at 1100 feet.

From the dam we drove west to Cleveland, Tenn. There was a *Goniobasis* reported many years ago from a "spring near Cleveland on road to Ducktown" which we wanted to look up, and we also wanted to investigate a small lake, also called "Country Club Lake", shown on the map near the town. We reached the lake just before dark and pitched camp in a grove of pines, the most ideal camping spot of the trip. Next morning we collected thoroughly in the lake and in its inlets and outlets, and took two fine *Goniobasis* and a beautifully-ribbed *Pleurocera* in great quantity. We found a few naiades also. Our catch here was particularly interesting as the lake drains into the Conasauga River. In fact, the ridge just back of the lake was the dividing line between the Conasauga and the Tennessee drainages. About noon we started a hunt for the elusive spring and found it without much trouble. There were a few of the *Goniobasis*, which we gathered, but did not linger long, for the water was icy cold and quite deep, and typhus was reported by the local hospital.

We now headed north for the Tennessee River at Loudon, which we reached at dusk. Here, at Sam's Island, Dr. C. C. Adams reported large numbers of *Io*, form *loudenensis*, and we were all on fire to diminish the local race. Next morning we borrowed a boat and went down river. At once we began to find *Strephobasis*, *Angitrema*, *Anculosa*, and several *Pleurocera* and *Ancylus*. But no *Io!* The Tennessee was still a little high, however, and we might have had better luck a month later. Finally, one of us did find the first *Io* of the trip, a fine, big *loudenensis*; and he at least will always remember the thrill he felt when he put his hand on those big spines. The thrill never wears off with *Io* even when we later collected several hundred in a day. Although we felt the bottom barehanded for many yards around that spot, a swift rocky shoal, we found no more. So a little after noon we drove north again to Lenoir City where our mail awaited us, stocked up on beans again, and drove northward till we struck the Clinch River at Riley Ferry, not many miles above the mouth. Here again we took only a lone *Io*, but a wonderful catch of naiades, many species; some of which were old friends. Although we saw ideal situations for *Io*, we could find none. Perhaps this is because we had not yet learned how to look for them, or because until the water is very low, they are quite dispersed. There is a trick to *Io* collecting. Later we found the most plentiful on fairly swift shoals, at the head of the riffles and at the end, just under the edges of rocks and smooth ledges, usually spire pointing down stream. Never in water over three feet deep, sometimes partly out of water. This is surely the queen of fresh-water shells.

During the next day we made a loop to the north and collected *Goniobasis* and *Physa* in Bear Creek, Poplar Creek, and several springs, coming out on the Clinch again at Emory River. We camped here and made extensive collections on the shoals, again bagging a lone *Io*, but many *Anculosa*, some knobbed, *Angitrema*, *Strephobasis*, etc. The *Anculosa* were particularly large and fine here.

Leaving Kingston, we struck west to run down another doubtful locality, Cane Creek and Toco Creek near the North Carolina line. We eventually found both of these and neither had any shells at that point. The rapid deforestation has changed the character of these creeks and given them a shifting, sandy bottom unfavorable to shells. Such is liable to be the fate of many other streams; and many others are being dammed for power, also changing the shell life. So they should be surveyed as soon as possible. In getting to these creeks, we followed the Little Tennessee and were amply repaid for our trouble by the rich collecting we got along it. At Coytee Shoals we found *Angitrema*, *Physa*, *Unios*, *Sphaerium*, and a very docollated *Pleurocera*, while a few miles farther at Morgantown Ferry we made one of our richest hauls—700 very large, fine *Campeloma*, probably *C. coarctatum* Lea, over 1000 specimens of a giant *Pleurocera* nearly two inches long, and the usual run of *Anculosa*, etc. Although we hunted carefully we found no trace of *Io* in this river, though conditions seemed ideal and some of the natives reported finding "spiny shells" on the shoals. Dr. C. C. Adams and his collectors reported the same lack of success. It would be interesting to follow this up and see if there are any in the head waters.

We examined the Tellico River at two places near Vonore and found a few *Pleurocera*, then headed west again for Sweetwater. Sweetwater Creek was one of the localities we wanted to check up on. It is polluted below the town, but we struck it just above the town, near the pumping station, and not only took a large series of *Pleuroceridae* from the creek itself, but also collected several kinds of *Goniobasis* from Cannon Spring and a small pond close by. This was a good haul.

We now continued north to Lenoir City again, collecting in Turkey Creek en route, and ferried over the Tennessee. We wanted to run into Cades Cove, and climb one of the peaks of the Great Smoky Mountains. This would enable us to cut Little River in many places. We certainly did survey the latter river thoroughly, for we took eight lots out of the river from points extending from near the mouth almost to the source. Our trip into Cades Cove and the hike up Thunderhead is one of the high lights of the summer. The road into the Cove, while in good shape, is surely the crookedest road in the world, for its length. Coming out we counted 226 turns in ten miles, many of the hairpin variety. This was Paul J. Adams' old stamping ground, and we left our machine in the barn of a friend of his, made up our packs and started on the hike up. These mountains are the highest east of Rockies, and though this one was only about 5,500 feet elevation (Thunder Head), it seemed much more before our tired legs reached the top. The going was made a little easier by picking up an occasional *Polygyra andrewsi* Binn. A dozen of these will fill your pack. Next to *P. chilhoweensis* Lewis, also found in these mountains, this is the queen of the *Polygyras*. Adams assured us that in a wetter season, land shells would have been much more plentiful. We saw far too few. We made the top in a cold fog, and hastened to seek shelter in the cabin of a cattle herder, old Tom Sparks, another friend of Adams. Early in the morning we tumbled out and went up the crown of the mountain just over the North Carolina line where we got a superb view of the rugged peaks. Then back for a hasty breakfast and the long pull down which seemed even harder on the legs than the hike up. Shortly after noon we got back to the car, packed up and pulled out. In packing we missed a bag containing two dozen eggs we had just bought and concluded the farmer's dogs must have gotten them. About a week later, Clench, after much complaint from the other two on the condition of his duffle bag, decided to clean it out and found the whole two dozen where the farmer had put them to be out of reach of his dogs! Curtain! We are still laughing at that.

We now turned north to Knoxville, stopping just once to collect several quarts of big *Pleurocera* from Nale Creek. We were in a hurry now because Clench felt an attack of his old malaria, contracted the previous summer, coming on. When we reached Adams' home in Knoxville, Clench was just able to wobble up to bed. Strong doses of quinine brought him around, however, and two days later he was out collecting again. We cannot express enough gratitude for the kindness and hospitality which the Rev. Mr. and Mrs. Adams showed us then and later.

As soon as Clench was over his attack we made a trip to the Holston River in company with the veteran collector Manley D. Barber, a resident of Knoxville and Rev. Mr. Adams. Mr. Barber promised us some *Io* on a shoals about twelve miles up the Holston, and was able to make good on his promise. For the first time we took *Io* in some quantity, nearly 200, as well as many other shells. These Ios were big spiny fellows, possibly typical form *spinosa*. On the way back we collected in Swan Pond Creek and had very good success. While waiting for Clench to get well, trips were made to the Tennessee River, with poor success, and to First Creek with wonderful luck. At the latter place, although the stream was quite cluttered with rubbish, the water was literally crowded with *Pleurocerids*. Never have we seen shells in such quantity. We filled up all our cans and buckets in a short time—we could have filled barrels—and then turned home. It would pay to make a complete survey of that stream, it is so rich, and we hope Adams will do it. Another bit of collecting we did in Knoxville was a search for land shells at Cherokee Bluffs and for *Helix nemoralis* Linn., introduced near the campus of the University of Tennessee. About 100 live specimens of the latter were carried to St. Louis by Remington and "planted" on the grounds of the Principia School.

On Aug. 30 we left Knoxville on the last leg of the trip. Our intention was to collect farther up on the Clench and the Powell and perhaps run into Virginia. We were going after *Io* in quantity now. Leaving Knoxville we collected in Second Creek, Knob Fork Creek and Beaver Creek, and struck the Clinch at Clinton. Here we collected a big series of *Campeloma* and *Pleurocera*. A few miles beyond we again hit the Clinch at the old Moore's Ferry and found the usual run of shells, in quantity, and about 100 *Io*, small, of the form *brevis*. We now headed for the mouth ef the Powell River where we wanted to get a big series, as there is some talk of a dam being built there. We found a few *Io* here in both the Clinch and Powell, as well as some fine *Strephobasis, Anculosa*, naiades, and the river forms of *Pleurocera*.

Another stop was made on the Clinch near Maynardsville,

Tenn. where we secured a fine catch of mussels. Then we made a detour to strike the Powell near Cumberland Gap, and had very good success there. For the first time we took over 100 *Io* which were nearly smooth. It would be hard to say what these shells are, as they intergrade, but they range between form *lyttonensis* and form *powellensis*. We kept zigzagging between the Powell and the Clinch here, until we turned northeast at Tazewell toward Kyle's Ford on the Clinch. A collection made in the Little Sycamore Creek was remarkable for having a few *Anculosa* in it, as well as some forms of *Pleurocerids* new to us. *Anculosa* were not usually found in small creeks. This day we made camp just before a pelting rain caught us. We saw the storm coming and stopped at the first handy spot, but quickly moved after the storm when we found we were camped close to a very dead calf! Once before we found we had camped near a departed mule. We favor a law in Tennessee requiring farmers to bury deceased animals.

We came out on the Clinch about six miles below Kyle's Ford, Tenn. and proceeded to collect all the way up to the ford. The river here is a series of shoals, quite clear, and we could see the Ios on the ledges. At Kyle's Ford they were particularly thick, as many as ten being brought up in one scoop of the hand. These Ios are from the type locality for the form *brevis*. It was thrilling work, though back-breaking, to grope for shells and keep a footing in the swift water. But our long-awaited dream of collecting *Io* had come true and we chalked off another aim achieved in conchology. We still have a few left. This was the last collecting we did, for our funds were almost exhausted and our time was about up too. Indeed, we had stretched the appropriations for two so as to cover expenses for three, and we considered that good managing. On Sept. 5, we drove back to Knoxville. All that remained now was to dispose of the car, pack our outfit and take the train back to Ann Harbor. The Rev. Mr. Adams kindly took the car off our hands and we were glad to know it was with friends. We feel very tender toward that car. On Sept. 8 we reached Ann Arbor again, having stopped in Detroit a few hours to report to Dr. Walker and spent a pleasant hour with him.

Our work, while very scattering, did approach completeness on the Little and Clinch Rivers—we collected at eleven stations on the latter river. As the upper reaches of the Clinch have been explored by Goodrich and Ortmann, the Clinch may be said to be thoroughly explored now. We estimated that we brought back over 250,000 specimens. Dr. Walker says no such volume of work has ever been done on the *Pleuroceridae* at one time. The whole work was really in the nature of a survey, and when the results are made clear, will furnish a basis for further work later. We want however, to give the Ios time to come out of their hiding places again. It is reported that the canned-bean industry had a big year! A wonderful trip, vagabonding for shells!

THE NAUTILUS.

Did You Know

that, though specimens of **Nautilus pompilius** preserved in the shell have been difficult to obtain in the past, you may now adorn your collection with a

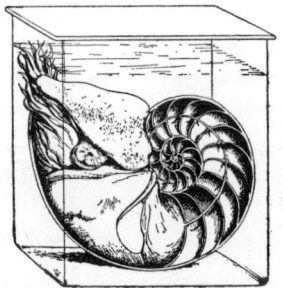

beautiful example of the "ship of pearl" nicely preserved, with shell sectioned showing chambers and siphuncle, and animal occupying the last chamber, mounted in exhibition jar, at very low rates. We have recently received a supply, and can furnish you specimens of various sizes, mounted or unmounted, cut or uncut. Also dry shells, sectioned or entire. Send for descriptive price list.

The above, however, is but one of the many thousand shells listed in our large new "*Catalogue of Mollusca*," 170 pages, over 200 cuts, with "*Supplement of Helicidæ*," etc , price 40 cents.

For anything in this line or in the departments of geology, mineralogy, palæontology. archæology, invertebrates, osteology, taxidermy, human anatomy, etc., you should communicate with

Ward's Natural Science Establishment,

A DAY ON THE CHICAGO DRAINAGE CANAL.

BY FRANK C. BAKER.

Frank Collins Baker was one of the malacological giants of the first quarter of this century. His early interests were in marine shells, but the circumstances of employment led him into freshwater mollusks. This resulted in several excellent tomes on the fluviatile fauna of Wisconsin, the Chicago area, and on the families of pond snails, Planorbidae and Lymnaeidae. He was interested in teaching young people, and produced a number of popular articles and books. He contributed 123 articles to The Nautilus.

F. C. Baker was born Dec. 14, 1867, in Warren, Rhode Island. Early in his life he was a Jesup Scholar in 1889 at the Academy of Natural Sciences of Philadelphia, and then worked briefly at Ward's Natural History Establishment in Rochester, N. Y. He was Curator of the Chicago Academy of Sciences from 1894 until 1915, and henceforth was Curator of the Museum of Natural History at the University of Illinois until 1939. His interests were in aquatic ecology and the taxonomy of freshwater pulmonate snails. Soon after becoming President of the American Malacological Union he died at 75 on May 7, 1942.

July 30th, the Chicago Academy of Sciences spent its annual field day on that wonderful engineering triumph, the Chicago Drainage Canal, and conchological results of the excursion may be of some interest to the readers of the NAUTILUS.

The day was all that could have been desired, the sun being more or less obscured by clouds, which made collecting more comfortable than under the boiling sun. The first stop was made at a point a few miles from the city, where the canal cut through the glacial clay or till. In a small stream by the side of the Santa Fé tracks, the conchologists picked up *Vivipara contectoides, Planorbis trivolvis, Sphærium stamineium* and *S. simile*, the first named species being very abundant.

The second stop was made just east of Summit, where the canal cut through blue till, in some places almost as hard as rock.[1] In one corner of the canal at this locality the bank and ground was fairly paved with minute shells perfectly preserved and of a whitish or chalky color. From this spot we collected *Bythinella nickliniana, Amnicola limosa, A. lustrica, Cincinnatis cincinnatiensis, Planorbis truncatus!, P. campanulatus, P. deflectus* and *Valvata tricarinata*, the last two species being represented by thousands of individuals. These mollusks are all referable to the Pleistocene deposits; *P. truncatus* was typical and very rare, as but one specimen was found. From the Desplaines River Mr. Woodruff collected *Alasmodonta complanata, A. deltoidea, Anodonta grandis, Lampsilis luteolus* and *Calyculina truncata*, the later very large.

At Willow Springs, which was the next station, I spent about three-quarters of an hour hunting for *Anodonta imbecilis*, but only succeeded in finding one half grown specimen. This is the only locality, so far as known, for this species in the Chicago area, and we had entertained high hopes of finding a "colony" of them, but such was not to be. The specimen collected was found in a soft, slimy, black mud, filled with broken bottles, tin cans, etc. Under an old bridge we found *Succinea retusa* very plentiful.

A long stop was made at Lemont to enable the palaeontologists to examine the many piles of limestone, which had been blasted from the canal, in search of Niagara fossils. Only a few were found, and those were very imperfect. Some brachiopods, a few mollusks, including several large *Cyrtolites amplicorne*, and an occasional Crinoid or trilobite was all that rewarded the geologists. The small boy got suddenly rich selling the common Niagara Calymene (*C. niagarensis*) at from five to twenty-five cents each, according to quality. No recent mollusks were found.

At Romeo, Dr. H. N. Lyon and myself walked half a mile north to the Desplaines River, and found a good collecting spot where the river ran over a bed of limestone arranged in ledges, and was quite

---

[1] See Leverett, Bull. 2, Geol. & N., 16 Surv., Chi. Acad. Sci., p. 49.

shallow. Here we found *Planorbis trivolvis, P. bicarinatus, Limnæa desidiosa* and *Goniobasis livescens*. Among the latter there were many which connected *livescens* with *depygis*, having well marked color bands and a purple tinted columella.

The last stop was made at Lockport where the train waited over an hour, and while the majority of the party studied the bear trap dam, the conchologists "pocketed" their cans and bottles and climbed (or fell) to a good sized creek (a branch of the Desplaines River). *Limnæa palustris* was here so abundant that it could be collected by the quart, and they were all large, fine specimens. Many specimens were very long and pointed and seemed to show a tendency toward *L. reflexa*. The stream was very rapid, and *Limnæa* and *Planorbis* seemed to be the only genera able to live in any numbers. *Physa* was abundant dead, but only three or four living specimens could be found. It decidedly prefers still water in this region. A single specimen of *L. palustris* was found in which the base had suffered some injury, and the aperture was thrown off to the right, leaving a wide and deep false umbilicus. We collected here *Limnæa palustris, L. caperata, L. humilis, Planorbis trivolvis, P. bicarinatus, Aplexa hypnorum* and *Physa heterostropha*.

Physa heterostropha at this locality shows a wide range of variation. Some are long and cylindrical, others broad and stumpy, and the spire runs from obtuse to pointed. The number of whorls was invariably the same. In this lot one could easily pick out such pseudo species as *gyrina, cylindrica, parva, oleacea* and *sayii*. The writer has recently tried Crosse and Fischer's suggestion in regard to specific characters in the form of the teeth on the radula, but thus far with a decidedly negative result.

The results of the field day, conchologically, may be summed up as follows: Pleistocene species 8, recent species 19. We carried home several quarts of mollusks.

TO SHELL COLLECTORS.

I am offering a very large and varied stock of Shells, Land Fresh Water and Marine, from all parts of the world, correctly named, and localities given. Specimens required will be sent on approbation to accredited parties.

SLOMAN ROUS, 929 De Kalb Ave., Brooklyn, N. Y.

SHELL COLLECTING ON THE MISSISSIPPI.

BY FRANK C. BAKER.

For a number of years it has been the custom of the Chicago Academy of Sciences to have a Field-day some time during the month of July and to spend the day investigating some notable or particularly interesting locality, from a zoölogical, botanical or geological standpoint. These excursions are not only attended by members of the Academy, but by the faculties and students of the Chicago University, the Northwestern University and kindred scholastic bodies.

Saturday, July 12th, was chosen as the field-day for 1902, which dawned bright and pleasant. About one hundred and fifty people, including many of the charming " co-eds " from the Zoölogical Department of the Chicago University, met at the Chicago, Milwaukee and St. Paul depot, from which the special train left at eight o'clock for Savanna, Illinois, our objective point. The ride consumed several hours and we arrived in sight of the Mississippi about noon.

Our first thought was for the " inner man," and we hastened in a body to the river bank, where we bargained with the boat renters and secured row-boats. No sooner were our bargains completed than we scrambled into our boats and rowed across the river toward a group of islands, where we ate our lunches.

The pull across the river was very interesting, especially to several of the " co-eds," who bravely volunteered to row one or two of the boats, for there was a seven-mile current which made this a matter of great exertion. The writer had never before seen the " Father of Waters," and he must confess that a peculiar feeling came over him as he rowed across the swiftly-flowing stream and thought of the many historic scenes which had taken place on or near this mighty river since De Soto first saw it. But the most interesting fact *to him* in connection with this river was that it afforded a home for more Unios than any other stream in the world.

As soon as lunch was out of the way we began a hunt for clams, and before the time arrived for the departure of our train we had accumulated several bushels, beside numerous examples of freshwater gastropods, such as Campeloma and Vivipara.

About a mile above Savanna we found several men engaged in "fishing" for clams, which they sold to the button-factories at Muscatine and other places in Iowa and Illinois. Their method of fishing was ingenious. A bar of iron (frequently a gas pipe) six or seven feet long is strung with four-pronged hooks, made of bent and twisted telegraph wire. The strings are about five inches apart and two or three hooks are attached to each string, making two or three rows of hooks attached to the bar. As many as forty hooks are frequently strung on one bar, the whole appliance being locally known as a "crowfoot" dredge or grapple. A piece of rope is tied near each end of the bar, forming a sort of bridle, and to this is fastened another rope, twenty-five or more feet in length, by which the dredge is pulled over the bottom of the river.

At first sight one would hardly suppose that with such an instrument a person would be able to gather very many clams, but the fishermen told us that several tons could be obtained with this apparatus in a comparatively short time. The clams are caught in this way: in many parts of the river the Unios lie packed by thousands, their shells half protruding from the mud and slightly gaping, as is natural with all these mollusks when at rest. As the fisherman pulls the dredge along the bottom over these Unio beds the prongs of the hooks become caught between the open valves of the shell, which immediately close and fasten themselves to the prong. A single haul may yield over one hundred shells caught in this way.

The inordinate collecting of shells for the button industry bids fair to exhaust the supply before many years have passed unless wise laws are enacted and enforced. Not only are many tons of these shells taken every year, but a large number are wilfully wasted by the fishermen. An example of this waste came under the notice of the writer on this occasion. Having failed to secure as many specimens as were wanted, a fisherman was asked if he knew a good place to gather clams. He replied that just above a large grain elevator some fishermen had dumped a boat-load on the shore. Not realizing fully what he meant, we walked to the spot indicated and there beheld a sight which made at least one of the party both glad and sad. Piled on the shore for a distance of a quarter of a mile were thousands upon thousands of clams, some alive, others with gaping shells and a few entirely devoid of the animal. Not less than twenty-five species were represented, many of them useless for the manufacture of buttons, but of great value to the conchologist of the future who may wish to study these species. The fishermen were either too lazy to throw them back into the water or else thought that if they threw them on the shore they would avoid catching them again on their hooks. Such wanton destruction as this, if not stopped, will soon exterminate many of the species. Those which were thus destroyed were comparatively thin shelled, such as Anodonta, Alasmidonta and Symphynota.

THE NAUTILUS.

VOL. IX.   JANUARY, 1896.   No. 9

TO CONCHOLOGISTS.

We regret the necessity which compels us to start the new year with an editorial of this nature; but it is nothing new to hear that the NAUTILUS must struggle for existence. This struggle has continued since the Jura, until now we have but a few species, three or four in the Indo-Pacific and one in the United States. Do you intend that the only recent NAUTILUS in North America shall become extinct? We are willing to supply the care, but not the entire environment. It needs feeding once a month in order to add another septum. You are asked to assist in this important function once a year; and when you see a slip of paper which reads, " Inclosed please find $1.00," it means that it has come your turn to " chip in." We hope that you will no longer neglect these little reminders. They mean that *your* subscription is due. We cannot wait until the end of the year—*we must have it in advance.*

Wishing you all a Happy New Year.

H. A. P. & C. W. J.

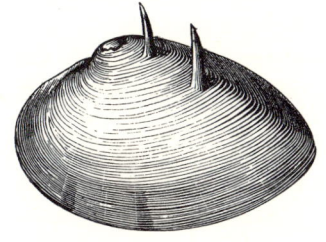

5

Passions of the Pearly Mussel

The rich freshwater clam fauna of America produced a breed of passionate conchologists, both amateur and professional. And little wonder, when one considers the many hundreds of gorgeous species of pearly mussels showing all manner of shapes, sculpturing features, and iridescent interiors.

Before the advent of massive pollution of our streams and rivers during the turn of the last century, it was possible to collect several hundred pounds of mussels, representing over a dozen species, on a single day's field trip in the Midwest.

Vol. v.   OCTOBER, 1891.   No. 6.

MOLLUSKS OF SPOON RIVER, ILL.

BY DR. W. S. STRODE, BERNADOTTE, ILL.

Dr. William S. Strode, a physician of Illinois wrote eight entertaining articles for The Nautilus *between the years of 1891 and 1898 on the joys of collecting pearly mussels. In 1898, his friend Berlin H. Wright, an avid unionid collector and former school teacher, named a Florida species of* Unio (strodeanus) *in his honor. Dr. Strode and his wife, and his collecting companion, Dr. Maguire, trained the latter's bird dog, Belva, to hunt submerged mussels and bring them ashore. In his article on Thompson's Lake Illinois, Dr. Strode tells of the great care he took of his catch of breathtaking* Anodonta suborbiculata, *each "as large as a common dinner plate" by covering his buggyload of precious clams with a wet blanket. His method of cleaning and preserving should interest today's collectors.*

Spoon river is a tributary of the Illinois. For a hundred miles from its junction with that stream its average width is about one hundred and fifty feet. It is a clear, swift-running stream, pursuing a sinuous course through a valley a half mile wide. Its banks are fringed by willows that here and there sweep the current in rhythmical response to every passing breeze. Overshadowing this border are silver-leafed maples, elms, and intertwining undergrowth, and beyond, towering above all like gigantic sentinels, stand the monarchs of the forest—giant sycamores.

No systematic study of the mollusks of this river has ever been made. Prof. Jno. Wolf, an aged naturalist of Canton, Ill., has made some researches, and probably knows more of the mollusks of the Illinois and Spoon Rivers than does any other living man, but he has written little of his discoveries.

Some of the Unios found, attain a size and perfection of form rarely equalled by shells of the same species found elsewhere. This perfection is due to the fact that each species finds in the variety of deep and shallow water, swift and sluggish currents, deposits of black mud, blue clay, sand, rock, and gravel, or a mixture of all these, the environment most suitable for perfect development.

Specimens of *U. multiplicatus* have been found over eight inches in length, and weighing three pounds. *Margaritana complanata* also grows very large, one specimen found two years ago being nine inches long.

A half a mile below the milldam at Bernadotte there is a noted mussel bed where for many years the fishermen have resorted for bait for their trout lines; here in a few minutes an ample supply of big fat mussels was to be had, and a catch of the toothsome channel cat assured. If an eel was desired the red meat of a *trigonus* was thought to be almost a sure means of luring the slippery *Anguillidæ*. Here within a space of two feet square I have taken at one time such species as *Unio plicatus, ventricosus gibbosus, asperrimus, pustulosus, tuberculatus, anodontoides*, and *Marg. rugosa*. A little higher up in deeper water and muddier banks *U. multiplicatus,* and *Marg. complanata* were plentiful. A little lower down, where there was much sand, the *U. occidens* and *anodontoides* could be found by tracing the path made by them in moving about. A half mile further down stream are great ledges of rocks that in places project far out over the water. This is a favorite resort for pic-nics, fishing parties, and experts at throwing the gig or fish-spear can sometimes obtain fine buffalo or catfish that are disporting under the shadows of these immense rocks.

At this picturesque point are to be found in considerable numbers, *U. trigonus, gracilis, pustulosus, tuberculatus,* and *lævissimus*. The latter, up to date, I have not succeeded in finding in any other location on the river.

Above the dam, where the water for a distance of five miles is from eight to fourteen feet deep, the *Ano. grandis* and the little *U. parvus* are the main shells. Here also are to be found many *Sphærium solidulum*, and *Paludina integra*. The *Physa heterostropha* and *Somatogyrus subglobosus* I find below in more shallow water.

On a large moss-covered rock I found at one time last fall large numbers of *Pleurocera Lewisii*, and in a few minutes gathered an oyster-can full. Visiting the locality again a few days later not one could I find, nor could I again locate them during the season.

The pearl craze struck this village last fall and wagon loads of the larger species were carried ashore and eagerly opened with the expectation of finding pearls that would at once enrich the possessor. The mussel bed before mentioned was almost annihilated. The final summing up showed about one hundred pearls of various sizes and colors. These were sent to Geo. F. Kunz, gem expert with Tiffany & Co., N. Y., who reported them of little or no value.

UNIO COLLECTING, BY DR. STRODE.

On October 1st I went to London Mills on Spoon River, about 40 miles up stream in the hope of finding *U. capax* and *U. aesopus*, but was disappointed. I was surprised to find *U. undulatus* Bar. superceeding *U. plicatus* and *U. multiplicatus*, so common lower down the stream. *M. complanata* Bar. was here in great numbers; *U. occidens* and *U. gibbosus* were also quite abundant.

On Nov. 10th, while on a picnicing expedition at Duncan Mills, 20 miles from the mouth of Spoon River, I observed on the opposite side of the stream a rocky ledge and beach below extending for quite a distance up and down the river.

The thought at once struck me that my giant *multiplicatus* might be once more found here. Accompanied by Dr. Maguire and our wives we crossed over and lost no time in getting into the water among the rocks. Almost the first shell brought up was one of these big fellows. They were here in company with scores of big *plicatus, ligamentinus, tuberculatus* and a dozen other species. In two hours' time we had found over fifty of the *multiplicatus,* one good *U. capax* and one *M. confragosa* four inches long. The doctor's bird dog Belva, partook of our enthusiasm and manifested a desire to also search for shells. After a little showing she understood how it was done, and it was amusing, indeed to see her with head submerged hunting a shell and then after securing it the air of importance assumed as she waddled ashore with it. We hope, the coming season, to make an expert collector of her.

One of the most pleasant and profitable collecting trips of the season was made in September at a place called "The Devil's Elbow," five miles below Havana on the Illinos River. At this place the south bank for nearly a half mile is a sand-bar, full of little bayous, and in these places was where we found the Unios. Prof. Hart, of the State Biological Station, who was one of the party, brought with one sweep of his dredge-net over fifty specimens, covering a dozen species. All of the following species were found plentiful, viz.: *U. plicatus, U. multiplicatus, U. alatus, U. gracilis, U. pustulosus, U. pustulatus, U. lachrymosus, U. anodontoides, U. gibbosus, U. ligamentinus, U. ebenus, U. ellipsis, U. solidus, U. donaciformis, U. cornutus, U. elegans, M. confragosa, M. rugosa, M. complanata.* A half-dozen *U. securis* were found, the first record of this species for the county.

Vol. XVI.   AUGUST, 1902.   No. 4.

COLLECTING UNIONIDÆ IN TEXAS AND LOUISIANA.

BY L. S. FRIERSON, FRIERSON, LA.

Lorraine Screven Frierson was a Louisiana cotton planter, born on August 7, 1861, in Lakeville, Louisiana. He developed a keen interest in the freshwater unionids. He became very proficient in his hobby and produced some major works in the field. He was a collecting companion of W. S. Strode and J. D. Mitchell of Texas. His extensive collection of pearly mussels is now at the University of Michigan, Ann Arbor. He produced 47 articles, mostly technical, for The Nautilus. *Frierson died in 1933 at the age of 72.*

In July, 1901, Dr. W. S. Strode, Mr. H. G. Askew, and the writer, took a trip through eastern Texas, collecting Unionidæ. Dr. Strode first took a "still hunt" on the Sabine river, at Loganport, where he duplicated the experience of the writer, the results of which have already been given the readers of NAUTILUS (xiii. 79). We met Mr. Askew at Sheperd, a small town northeast of Houston, and in close proximity to "Big creek," and Trinity river.

From Big creek we obtained a few *Lampsilis lienosus*. This shell had never before been obtained so far west, nor had it been listed as a Texas species by Mr. Singley. The Trinity river, though shallow at this time and place, was swift, with a sandy bottom, a combination not favorable to unio life, and we had therefore poor luck. We obtained some magnificent *Quadrula pauciplicata*; big, glossy, black and nearly devoid of plications. They were otherwise interesting on account of the females being gravid, an unusual condition in this group. It is a true Quadrula in this respect. Some very fine *Q. trapezoides* were also taken. They were remarkably compressed, and some of them were likewise gravid. They bore their young (or eggs) in all four gills. This we believe has never before been noted, and effectually places this species in the genus *Quadrula*, as defined by Mr. C. T. Simpson, who placed it here without having the advantage of seeing a gravid female. We captured a trio of *L. amphichænus*, which extends both the habitat and size of this remarkable species, one of them being $5\frac{1}{4}$ inches in length. (The writer has since obtained a dead shell from the upper Brazos river.) A fine series of shells were found which are in my cabinet as yet unnamed. They seem to be a perfect connecting link between *Q. aurea, houstonensis*, and *pustulata*. We were fortunate enough to find a couple of *Q. chunii, Lea*. This is the river in which the types were obtained and the specimens were typical in every respect. This shell is a very rare species, and one sadly abused. Whenever a uniologist gets a shell belonging to the group headed by *Q. trigona*, and about whose name he is in doubt, he at once dubs it *Q. chunii*. I may be rather harsh on my brother uniologists, but these two shells are my only *chunii* to date.

The next day we were at Nacogdoches, Texas. Here we saw the celebrated "Stone Fort," an ancient structure over whose walls the flags of *seven* governments have floated. How many of my readers must plead guilty, as I did, of never having heard of the Republic of Fredonia?[1] The full history of this structure was given us by Mr.

---

[1] In April, 1825, Hayden Edwards made a contract with the government of Mexico for the introduction of 800 families into Texas. They were to settle in the neighborhood of Nacogdoches, and be provided with lands under the general colonization law. The location proved unfortunate. Nacogdoches had been settled many years, partly by Mexicans and partly by a roving class of people who had a prejudice against the Anglo-Americans. When the colonists selected their lands and commenced improving, some older claimant would appear. The courts were appealed to, but would invariably decide in favor of

Askew, who is as loyal a son of Texas as ever drew breath. It is a shame to the town that this fort has been recently torn down and replaced by a sordid brick store. As soon as we had breakfasted, we went to the La Nana creek, where we obtained the new species, *Q. lananensis* recently described. We also obtained a number of the most deeply corrugated *Q. laticostata* we have yet seen. A solitary *Obovaria castanea* was taken. Numbers of *Tritogonia tuberculata* were found, but much to our disappointment, not a single gravid female was noted. (This species has not as yet been observed in that condition.) In this creek we obtained some *L. nigerrimus* and *Strophitus edentulus*, neither of which was listed by Mr. Singley. While cleaning up our catch in the hotel yard, we were joined by an intelligent-looking party who gravely asked if the "fossils" we were cleaning belonged to the Devonian formation! I shall never forget the guileless look of the doctor, as he gravely replied that they *did*.

By high noon next day we were at Rockland on the Neches river; we had taken our dinner, and by 5 p. m. were loaded with all the unios we wanted. This place is the metropolis of *Q. askewii*, of which some examples require a "Philadelphia lawyer" to differentiate from *Q. beadleanus*. The unios of this river are precisely the same as in the Sabine river. We obtained some *bona-fide Q. nodifera*, a species of the validity of which we had had doubts, but these are

the Mexican constituents. These conditions continued until finally (1826) the Mexican governor of the province decreed the annulment of the contract and the expulsion of Edwards and his brother from the territory. But Edwards had expended several thousand dollars in this enterprise, and his colonists too had expended considerable in building their homes. The Indians (principally Cherokees) also had settled near-by under the provisions of the colonization laws, and being greatly dissatisfied, allied themselves with the Edwards colonists, who, assuming the name of Fredonians, declared their independence of Mexico. They proceeded at once to organize a legislative committee composed of eight Americans and five Indians. Learning that Col. Bean was preparing to resist their movements, they took possession of the old stone fort. Norris, the deposed Mexican Alcalde, collected some friends and on Jan. 4, 1827, entered the town; they were attacked by the Americans and Indians and driven off with a loss of one killed and several wounded. The Fredonians were sadly disappointed in not receiving the co-operation of the Austin colonists, who joined the 200 soldiers sent from San Antonio to suppress the infant republic. Seeing the hopelessness of maintaining the Replubic of Fredonia, Major Edwards and his forces retired across the Sabine into the United States and disbanded.

We are indebted to Mr. Askew for the above notes from Thrall's History of Texas.—EDITORS.

now forever laid aside. From Rockland we then took flight towards Lake Charles (Louisiana). En route we were compelled to stop over at Beaumont, Texas; while there we were fortunate enough to witness the striking of oil by one of the wonderful "gushers" of that place. It was a grand sight, the memory of which will never leave us. Lake Charles we found to be a shallow expansion of the Calcasieu river, about two miles wide, with sandy bottom, and covered by floating masses of the "Water Hyacinth," acres and acres of them. Calcasieu river is an extraordinary stream; for fifty miles it is sixty feet deep and a quarter of a mile wide, with no current excepting after rains, and not a shoal or sand-bar. The salt water comes up 40 or 50 miles during storms, and kills most of the fresh-water shells.

INHABITANTS OF A NATURAL AQUARIUM.

BY L. S. FRIERSON.

Red River having become choked by vast accumulations of drift-logs in the vicinity of Shreveport, Louisiana, carried its waters to the Gulf through many side channels, which soon became possessed of high banks (as had the main river), and the lower lands between these channels acquired local names, some as "lakes," others as "bayous".

The drift however, having been cleaned out by the U. S. Government, and the side channels dammed at their heads, most of the lands constituting the Valley of the Red River are now in cultivation, even some of the former navigable lakes being cultivated.

When first explored by the writer, Bayou Pierre even at low water stage was a fairly large stream, and entitled to the name of "river".

The bed of this stream was swarming with millions of mussel shells, comprising nineteen species.

The creeks emptying into Bayou Pierre in this vicinity contained water of very different kind from that of Red River, the latter being heavily charged with gypsum, lime, and in low water stages even salt could be noticed as one of its flavoring materials. But the creek affluents of the river carry quite "soft" waters, and this difference, if not the cause, is at least correlated with a quite different mussel fauna. *Anondonta grandis* is the single species common to both creek and river.

When the head of Bayou Pierre was dammed across, there ensued of course a tremendous mortality in the naiad population, hundreds of acres of hitherto living waters becoming dry lands. Gradually the Bayou Pierre has become converted from a stream containing the hard water of Red River to one containing the soft waters of the local creeks, and in fact is now only a large creek, going dry during droughts, except in local pools.

Between Red River and Bayou Pierre a low valley was for long known as Brown Lake, but which now is in rapid process of being put into a high state of cultivation.

A rail road, and a hard surfaced public road now traverse its former site. Alongside of the latter a ditch was dug, five feet wide and two feet deep and in the lower part of the lake site this ditch holds water for some time after rains, during which the ditch communicates with Bayou Pierre situated about a mile away.

In such flood times, fish run up these temporary streams, seeking pools in which to lay their eggs, and as these are ofttimes infested with *glochidia* the bottom of the ditch abovementioned becomes sown with young mussels.

It has so happened that the past two years have been unusually wet, and the rains have been quite equably distributed during the year, and hence the ditch in question, in its lower portion of about two hundred feet in length, has been continuously more or less full of water, until the present autumn (1922).

The writer had occasion to walk down this dry ditch and somewhat to his astonishment found hundreds of mussel shells on the bottom, some of which being collected proved of much interest.

A single *Anodonta grandis* was found, almost five inches long, showing a quite rapid growth, for it is impossible that this shell is more than *thirty months* old; most likely its age is only eighteen months.

The most interesting cases however are of the two following shells. The writer, in Nautilus, 1903, showed that *Unio tetralasnus* Say, with its several synonyms and the *Unio declivis* Say with its synonym *geometricus* Lea, were entirely distinct species, differing in shape, size, color of nacre and habitats.

This has been strikingly proven true by the changes in the local conditions outlined above. In the dried bed of Brown Lake, great numbers of old dead shells of the *A. geometricus* can still be picked up; but in the many years of personal collecting done by the writer, no specimen of *tetralasmus* has ever been found in any Red River water. How interesting it was, then to find that the bottom of the ditch mentioned is teeming with typical *tetralasmus*, and not a single *geometricus* exists, I am sure, in this vicinity.

The latter form is universally held by all writers, including Lea himself, to be a local form of *Unio declivis* Say. The single exception to this reference was Simpson, who in his Catalogue of 1914, cites the figure of *geometricus* given in Nautilus, 1903, Plate III, as being *camptodon* Say!

The population of this local aquarium however contained another surprise. For many years the writer has tried to prove by concrete material, what he was convinced to be true, that *Unio haleianus* Lea was merely an individual variant of *texasensis*; but no material had ever been obtained which could *prove* this *intuition*. Along with the *tetralasmus* in this ditch, the writer found hundreds of *texasensis*, and to his delight, a specimen of extra large size proved to be *typical haleianus!*

Although the bed of this little pond has been dry for the past two months, all of the *tetralasmus* are still living, and quite a number of the *texasensis* are also alive, but the majority are recently dead.

Notwithstanding that this pool of water was very seldom more than one foot deep, it seems to have been an almost ideal habitat for the three species mentioned, so long as the rains lasted.

One of the conditions which rendered this pool an almost *optimum* locality is the fact that being situated in an open commons, there is no shade, not even of weeds, to obstruct the sunshine.

It may not be known to every reader that the paucity of Naiades in the Tropics is thought by those who have collected in those regions to be largely due to the dense shade covering all but the larger streams.

The exploration of this ditch however furnished still another item of interest. As the pool dried up, the exposed *texasensis* began to die, and their valves gaping, the exposed contents were eaten by birds, and the latter not being content with their daily dead, in several cases undertook to expedite the process by pecking holes through their valves. With such force was this done that, in every case noted, *both valves* were punctured at once. Whether this action of these birds is due to *instinct* or to *reason*, the writer being strictly a Naiadologist leaves it to other better equipped observers to decide; merely remarking that this process has been previously observed, and the pecked shells in the writers cabinet now number three, from widely separate localities.

UNIONIDÆ IN A TUNNEL.—I am interested in two examples of *Margaritana margaritifera* var. *falcata*, taken in a water tunnel near Santa Cruz, in this state (California), 700 feet from the mouth of the tunnel, and 300 feet underground. They differ from the normal specimens in being both unusually large and thin, the nacre being very richly colored.—FRED L. BUTTON.

New Check List of North American Naiades.

By B. H. WRIGHT and BRYANT WALKER.

Arranged according to Simpson's Synopsis, including Synonymy, and brought down to May 1, 1902. 19 pages.

Indispensable for arranging collections and making exchanges.

Price, 2 for 25 cents. Apply to

B. H. WRIGHT, Penn Yan, N. Y.

MOLLUSKS OF THOMPSON'S LAKE, ILLINOIS.

BY W. S. STRODE, M. D., BERNADOTTE, ILL.

The beautiful *Anodonta suborbiculata* of Say has a sparse distribution and is rarely found in considerable numbers.

I know of but one locality in Illinois where it is to be found in abundance. This place is a still beautiful lake, five miles long by one in breadth, with an average depth of from five to eight feet; the bottom a mixture of black mud and sand; the shores and a hundred acres or so at each end covered with a growth of pond Lilies.

For a half century this lake has been a great fishing resort. With seines five hundred yards long, trammel and funnel nets, hook and line, spears, etc., immense quantities of fish are annually taken from its waters; great Buffalo, Cat-fish, Shovel-fish, Jack-salmon and a half dozen kinds of Sunfish, Bass, Pike and Pickerel.

State Geologist Worthen (deceased) seems to have been the only naturalist who discovered the conchological richness of the lake, and he kept the discovery to himself, collecting large quantities of the Ano. suborbiculata Say and corpulenta Cpr. and sending them to collectors and museums all over the world.

In the summer of 1890 I made a careful search for the Unionidæ and found it containing but four species: Unio anodontoides Lea, and parvus Bar., and Ano. suborbiculata Say, and corpulenta Cpr.

But the abundance of the two Anodontas make up for the lack of species. In some places the bottom of the lake seemed to be literally paved with the suborbiculata. With a six-tined potato-digger I would sometimes bring up five or six at a haul; and if the fishermen happened to be making a draw with the great seine, a half barrel of them would sometimes be drawn out at once,—many of them great beautiful adult shells nearly as large as a common dinner plate, the epidermis all intact, the stillness of the water and freedom from acid causing but little erosion. The younger shells in their beautiful iridesence, seem to have caught the tints reflected from the green woods, the blue sky and sparkling stars.

The other Anodonta, the corpulenta was not so plentiful in the deeper water that the suborbiculata seemed to prefer, but nearer the shores in shallow water, more or less shaded by the broad leaves of the water lily, many of them could be found. The umbones of this mussel, as found in this lake, more nearly approach perfection than in any other species.

Associated with this mollusk, among the water lilies, were great numbers of *Vivipara contectoides* Binney and *intertexta* Say and also more or less of the *Physa heterostropha* Say.

In collecting and handling these fragile shells much care must be taken as they break as easily as egg shells. When removed from the water I would pile them up in one end of the boat, and cover them up from the sun with a wet blanket. When transferred to my buggy (for I had to drive twenty miles to Bernadotte) I would first line the bottom of the bed with wet grass, on which I arranged the mussels and then again covered them up well with the wet blankets. On reaching home they were at once transferred to a large tub containing water. They must be cleaned without the use of hot water and immediately given a good bath of glycerine, and then kept in a cool place.

A GOOD CHANCE FOR COLLECTORS IN UNIONIDÆ.

Mr. Berlin H. Wright of Penn Yan, N. Y. will spend most of the winter collecting in the waters of Georgia and wishes some congenial company. Anyone favorably inclined may learn something to their advantage by addressing him as above.

WANTED. Works on Land Shells, and rare North American and Foreign Helices. OFFERED, *Mesodon dentiferus*, *Sayii*, *Acanthinula harpa*, etc.—*A. W. Hanham, Bank of British North America, Quebec, Canada.*

To EXCHANGE in large or small quantities. *Anodonta suborbiculata* Say and *Anodonta corpulenta* Cooper, also about 40 species of the fine Spoon river Uniones and univalves.—*Dr. W. S. Strode, Lewistown, Ill.*

EXCHANGE. *Unios heterodon* Lea, *lanceolatus* Lea, *Tappanianus* Lea, *Anodonta Williamsii* Lea and *Marg. undulata* Say (pink var.) for other *Unios* not in my collection.—*W. T. Farrer, Orange, Va.*

THE MANUFACTURE OF PEARL BUTTONS FROM FRESH-WATER MUSSELS.

In the manufacture of pearl buttons the centre of activity has shifted from the China Sea to the river towns of the Mississippi. Altogether unknown in this region a dozen years ago, this industry has grown to such proportions that it now employs the services of thousands of people, and the output has become so great that it materially affects the button market of the world.

About twelve years ago a German buttonmaker named Boeple wandered into Muscatine from the old country. He saw for the first time the mussel shells of the Mississippi river. He examined them closely and expressed the opinion that they were good material for buttons. Up to this time fresh-water shells were considered unsuitable for any such use, and authorities on the subject were naturally skeptical in regard to Boeple's opinion of their usefulness. He persisted in claiming that the "niggerhead" mussel from the waters of the Mississippi river would make, if properly handled and finished, the finest pearl buttons yet produced. He took some specimens to the factories at Waterbury, Conn., and after considerable experimenting one concern there determined that with some changes in their machinery the shell of the strange mussel from the "Father of Waters" would make a button to compete with the best of those from other parts of the world.

First one concern and then another began to use the Mississippi shell, until the foreign one was almost abandoned. In the beginning the shells were shipped east in the rough and prepared for use after their arrival there, but the freight rates were so high that one enterprising firm soon shipped that part of its machinery which makes the "blanks" out to Muscatine, and, what generally results when some pioneer leads the way to a good thing, others soon profited by the example and came also. The industry has spread both up and down the river, until almost every town of any importance, from St. Paul, Minn., to Alton, Ill., is now engaged in some form of the industry.

The manner of catching the mussels is interesting. A fisherman equips himself with what is known to the clan as a "John boat." This is a flatboat on the order of a scow, about 20 feet long and $3\frac{1}{2}$

feet wide. Upon the inside of the boat are placed eight uprights, which are between three and four feet high and have crotched tops. Four of the uprights are placed on each side of the boat, at just enough distance apart to accommodate the four 10-foot pieces of inch gaspipe that rest upon them. To each of the gaspipes are attached 20 four-foot stagons, similar to those used on an ordinary trout line, and each stagon has four hooks, with four prongs.

The fisherman goes out in his "John boat" with as much confidence as if it were the finest craft afloat. Once in the stream, he casts his gas-pipes, one by one. As the hooks drag along the bottom of the river they come in contact with the open shells of the mussels, which immediately close up on them. Thus attached, they are brought to the surface and taken off. The distance the hooks are dragged each time depends altogether on the thickness of the bed, and varies from three boat-lengths to an eighth of a mile.

The rivers of Arkansas are said to be so thick with mussel beds that they crop out of the water when it is low. The men put on rubber boots and shovel the shells into the boats. In the Upper Mississippi district, shells are quoted in car-lots, ranging from 15 to 30 tons in weight, but the Arkansas dealers have astounded everybody in the business by sending out quotations on 500-ton lots and promptly filling all orders sent them. The men sell the mussels to the button factory operators at so much per 100 pounds. The wages they make depend upon their diligence and the luck they meet with in getting in a thick bed, but range from $1.50 to $5 per day. There is one big mussel bed near Canton, Mo., about eight miles in length.

The process of making the shells into buttons is interesting. The shells are first cut up into blanks the exact size the buttons are to be; then they go to the grinder, a machine which grinds the black back off of them; after that to the facing machine, which cuts the face on them; next to the backer, which bevels the back; then the drill, which puts in the eye-holes; from here they go to the polishing room, where the glossy finish is put upon them; after that they are sorted, put on cards and boxed up.

There is an added interest in the business of mussel fishing on account of the likelihood of finding pearls. It is not an uncommon thing for a fisherman to find a pearl valued at $100, and one lucky fellow found a beauty which sold for $5,000. Every follower of the business has a little bottle filled with specimens, which eventually find their way to the market.—*Phila. Record.*

THE KENTUCKY RIVER AND ITS MUSSEL RESOURCES. By Ernest Danglade. Bureau of Fisheries Document no. 934. 1922. This brief article gives some account of the river, with a list of 40 species of mussels. The following summary may be of interest:

"The Kentucky River is approximately 400 miles long and contains many valuable mussel beds. In the upper reaches of the stream these number about two per mile of channel. They have well-defined and characteristic locations easily marked.

"The upper Kentucky River is practically an unknown and unworked mussel-bearing stream and contains an abundance of mussels of commercial value, possessing good nacre and texture. Of these the mucket constitutes about 90 per cent. This shell has a desirable color, texture, and uniformity of thickness throughout. It appears probable, therefore, that this stream may be particularly useful in the near future as a source of remunerative employment for the mussel fisherman and of desirable raw material for the button manufacturer. The pearls of this river, as a by-product of mussel fishing, are of small consequence, both in the quantity and quality of the pieces found.

"The railroad and steamboat shipping facilities of the main river, of North Fork, and of lower Middle Fork are satisfactory. On the upper Middle Fork and on South Fork there are no railroad or steamboat connections and shipments must be handled by small boats. In particularly dry seasons of the year transportation must be made by hauling over rough roads. This is especially true of the South Fork.

"The method of shell fishing in the Kentucky River is limited principally to hand picking or to the use of the shellfork. A stiff bottom in which the mussels bury themselves deeply makes implements commonly used elsewhere in shelling, useless in this river.

"Of 40 species of mussels observed as indigenous to the river, 22 are commercially usable, but only 9 are of relative importance. This number includes as the most common shells suitable for button manufacture: the mucket, the pocketbook, the pimplebacks, the pistol-grip, the long niggerheads, the maple leaf and the fat mucket."

Court Rules Mussel is not a Wild Animal

Chief Justice Oliver Wendell Holmes (1841-1935) must have been amused at this case, and probably agreed to take it on because of his father's interest in shells. Holmes, the elder, was a physician, Professor of Anatomy at Dartmouth and Harvard, and a great poet, famed among conchologists for "The Chambered Nautilus", (1836), and known among American patriots for his poem, "Old Ironsides". The latter whipped up enough sentiment for the ultimate preservation of the U.S.S. Constitution (1830). Dr. Holmes died in Boston on Oct. 7, 1894.

Court Rules Mussel is not a Wild Animal. The fresh-water mussel of the Mississippi valley, from whose shell is manufactured pearl buttons, is the property of the owner of the land through which flow the non-navigable streams of its habitat, the Supreme Court held.

The controversy attracted wide attention because of its importance in the button industry, and brought into court such questions as whether the mussel was a "wild" animal. Justice Holmes, in deciding the case, said mussels should not be classed with wild birds and fish which move beyond the jurisdiction of the owner of the land.

The lower court held that fishermen taking mussels from non-navigable streams could be prosecuted as trespassers, and the owner of the land through which the stream flowed could recover damages from whose who purchased shells from trespassers. This judgment was affirmed by the Supreme Court in principle.—*Boston Herald.*

The largest fresh-water pearl on record was found at Genoa, Wisconsin, by seventeen-year-old Frank Hastings while he was fishing. It weighs 185 grains and is pure white. It measures $\frac{15}{16}$ of an inch in diameter. A local dealer bought the pearl, just as it was when it came from the shell, for $2,675.—*Cleveland Leader.*

The Chambered Nautilus.

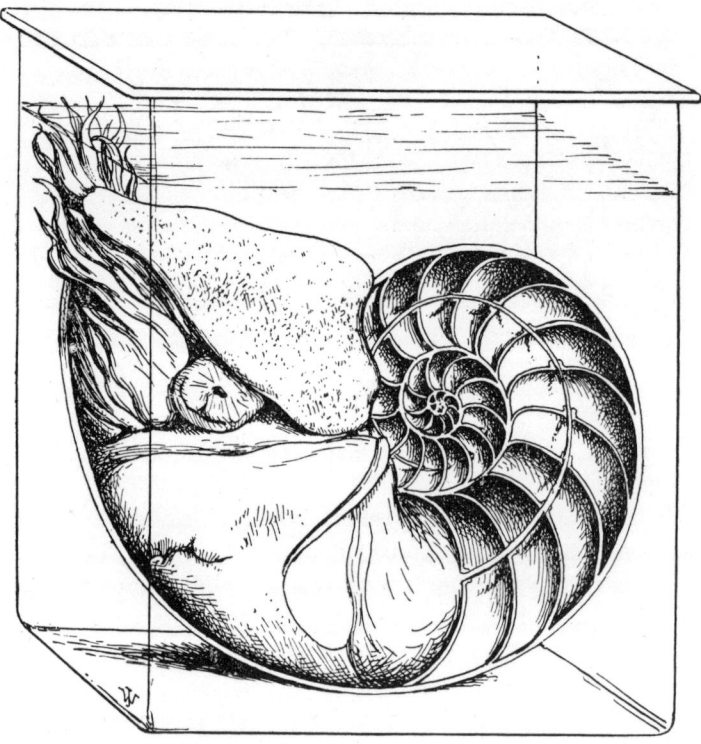

NAUTILUS POMPILIUS, Linn.   Western Polynesia.

Specimens in alcohol of this rare animal, with shell bisected,
    mounted in glass jar, as shown above, ... $12.00
Sent moist without jar 10.00
Non-bisected 10 00
Non- bisected, without jar 8.00
    Small-sized, $3.00 less than above.

Write for special circular to

Ward's Natural Science Establishment,

Collecting Abroad

Charles Hedley, the distinguished Australian malacologist, naturalist and explorer, was a personal friend of Henry A. Pilsbry, editor of The Nautilus *from 1889 to 1957. Hedley frequently sent interesting articles to his Philadelphia friend, among which we include one of the first accounts of helmet diving for shells. Hedley was born in England in 1862, and spent much of his youth in the south of France and in Switzerland. He went to Australia as a young man, and became a naturalist, first at the Queensland Museum and later at the Australian Museum in Sydney. He explored New Guinea in the 1890's and made field investigations at Funafuti Atoll in 1896. He may be con-*

sidered the father of Australian malacology, since he contributed more scientific information on the subject than any other person during the nineteenth century. Hedley was also an accomplished writer of popular articles. He died on September 14, 1925, at the age of 63, at his residence, "Nukulailai" in Sydney, of pneumonia.

A SHELL HUNT FORTY FEET UNDER THE SEA.

BY C. HEDLEY, SYDNEY, AUSTRALIA.

To widen the fairway of Port Jackson (Australia), a submarine reef is being removed. An opportunity of going down with the divers employed thereon was kindly offered to myself and a scientific friend by the officer in charge of the operations. So tempting an invitation was, of course, accepted with delight. Often in imagination had we wandered on the ocean floor, peering into ghastly wrecks of ships sunk long ago, fighting with some huge shark or monstrous octopus, and gathering treasures of science or heaps of gold. Now our dreams were to come true and we were indeed to tread that fairy-land. We might not have the luck of the mariner in the song who

> " Fell overboard in a gale,
> And found down below where the seaweeds grow,
> Such a lovely maid with a tail,"

but we should certainly pluck strange growths at the bottom of the sea as one might pick flowers in a meadow.

A trim launch sped with us from Circular Quay down the famous Sydney Harbor, past bay after bay, some lined with wharves and shipping and some with trees growing to the water's edge, by rocks and white sandy beaches, past point and headland gay with villas and gardens, or sombre with eucalypt forest. So familiar was the scene to us, that we smoked and chatted, unmindful of its beauties, till we reached a flotilla of punts and barges moored near the Heads.

After a cup of tea with the overseer, we prepared for our descent by divesting ourselves of boots, coat, vest and collar. A couple of laborers officiated as my *valets de chambre*, wrapping me first in thick flannel socks, trousers and jacket, and then in a canvas overall garment which left only the head and hands uncovered. The hands being left bare, the sleeves were secured at the wrists by rubber cuffs and bracelets. My feet were thrust into a pair of enormous boots, each sole of which was weighted with 25 pounds of lead. Bending my head, two men placed over it a huge diver's helmet and screwed it into a brass collar of the canvas dress. My costume completed by slinging on chest and back two large metal weights, I was told to rise. Thus encumbered, it was no slight exertion to get up, take three steps to the ladder, and descend into the water knee deep. There I halted while my signal cord was belted round my waist; my air-tube, which reminded me of a garden hose, was screwed to my helmet and the pump commenced to force air through it. Finally an attendant screwed a plate-glass front, the size of a saucer, into my helmet; from the inside, this last operation resembled the closing of a coffin-lid. Some one tapped my helmet twice, the submarine signal for "all's well," and I started.

Stepping off the bottom rung of the short ladder, down I went, till the keel of the barge loomed up, rose and passed me—down, down into the green sea water, watching the silvery bubbles stream upward—down, down, down, as the water darkened. That sensation of gliding down into an emerald abyss, was the weirdest, dreamiest thing I ever felt. Then so gently did I alight, that I merely noticed that I had ceased to fall. At my feet I saw rock and sand and seaweed; looking up, I saw a monster in a helmet with two ropes leading away up to where the sky ought to be. The monster's face showed through his little window as a big, fair moustache and a pair of kindly blue eyes. Fetching out of a capacious trouser pocket a small school slate, he wrote, "How do you feel? Shall we go on?" and held it up. Taking his slate, I wrote, First rate; go on." He read the message, gravely rubbed the slate clean with his finger, pocketed it, and held out his hand. I grasped it and we started for a walk at the bottom of the sea.

Then I noticed a pain in my ears; the compressed air was hurting me. To cure it, I went through the motion of swallowing once or twice. Feeling more comfortable, I "began to take notice," as they say of the babies. The light was bright enough to see small things plainly twenty feet away, but the water strangely magnified familiar objects. A shoal of little fish passed us, swimming under our arms and between our legs in the most ridiculous way. I tried to take one with my hand, but it deftly turned and avoided my grasp. The guide, seeing my attempt, pinned one to the ground with an iron rod he carried, and handed it to me; another he stabbed and caught as it swam by. Before we had gone far I had lost all sense of time, space or direction, and became too confused to know whether I had travelled east or west, ten yards or a hundred, in ten minutes or half an hour. A queer sensation was that of having escaped from the law of gravity; it seemed just as easy to walk up as down a cliff—we usually walked on our toes, sloping from the ground at an angle of forty to sixty degrees. When too much air is pumped down, the submarine pedestrian is unduly buoyant, and his aims to clutch a shell from the ground must be comically like the dodging and staggering of a drunken man.

A little dell lay before us choked with rank seaweed, through which we strode waist-deep like plunging into a tangle of fern in some damp valley on the land. My guide reached out, picked something off a broad frond, and handed it to me. It was a *Doris*, a lovely creature, whose like I never saw in books, striped with purple on a milk-white ground. It began to crawl over my fingers quite unconcernedly. I clapped my hands and tried dumbly to express my delight by patting my companion's big fist. He replied by offering me the slate, on which I wrote, "Very good; put him in the bottle." Rubbing out my words, he wrote, "Send down the bottle," tied the slate to the rope and jerked the latter four times. Away went rope and slate to the regions above. In response to an answering signal, the slack was hauled in and my collecting-jar descended tied to the rope. In turn, we tried in vain to open it. Although our correspondent above had filled the bottle with water, the pressure at our depth so sealed it that we could not raise the stopper. With a message on the slate, "Open this bottle and send it down open," we sent the jar aloft. When is was lowered to us the second time, I found that my *Doris* had slipped unobserved through my fingers, and so I lost a possible new species, the rarest treasure I was to see that day.

Continuing our travels in the dim water-world, we passed through a field of sponges. Not the brown, round masses of the bath-room, but radiant growths of scarlet (*Raphyrus hixoni* and *Halicondria rubra*) and purple, here and there great open oscula, tempting one to poke in a mischievous finger. Some grew in tufts like moss, some expanded like a dainty vase (*Phyllosiphonia caliciformis*), some forked like branches of trees and some spread like a lady's fan. One abundant species, about the size and shape of an orange, was pure ice-white, studded with golden dots that almost glittered (*Leucondra* sp.). Of all these we gathered what we could, pricking our hands sore with sponge spicules as we worked. When, on the morrow, our ravished beauties lay dead on a table in the museum, they had faded sadly from their pristine splendor. Among the sponges grew purple *Boltenia pachydermatina*, a pear-shaped head upon a slender stalk, like tulips in an earthly garden.

For a surprise, the diver held up before my face and pressed an Aplysia. From it flowed a violet stream which stained the water for two feet around, hiding hand and mollusk in the cloud. One of my last captures was an exquisite nudibranch, which swarmed on the broad fucus blades. In hue it was the blue of a summer sky, flecked with blood-red dots and stripes. I had now grown weary; not of searching for wonders, but of supporting the heavy diving armor, and was content to be drawn up again to the world of air and sunshine, which I had quitted three-quarters of an hour before.

On reflection, I found the reward of my under-water foray to be, not a hoard of specimens, but a better appreciation of the circumstances under which marine life exists. Our party of four had only observed, dead or alive, *Chamostrea albida, Vola fumata, Trigonia lamarcki, Struthiolaria scutulata, Drillia oweni, Cassis pyrum, Cypræa xanthodon, Astralium tentoriforme, Ranella leucostoma, Aplysia keraudreni, Chromodoris bennetti,* and two undetermined *Doris*. Molluscan life seemed, on the spot I explored, to be less plentiful than at low-tide mark. Perhaps, however, the difficulties under which I labored as a beginner in the art of diving, impeded me from finding what was really there. After seeing the rough sea floor, one wonders that a dredge should capture as much as it does. A rich harvest probably awaits a conchologist who should seriously practice diving as a means of collecting.

A DAY ON THE GREAT BARRIER REEF

BY CHARLES HEDLEY.

"All hands on deck!" Ugh! "The billy's boiled!" Groan. "The tide is falling fast!" That fetched your conchologist. Sleepily he crawled out and reached for his mug of hot tea. Balancing himself on the combings he looked abroad.

Far in the west the jungle-clad mountains of tropical Queensland loomed soft and blue. Between lay a purple sea which in the near distance suddenly changed to the vivid green of shoal water. To windward a beach of coral sand showed white and bright under the dense foliage which smothered a long, low island. Beyond it a line of foaming breakers stretched to the sky-line. Back against the white surf there rose the bristling fangs of the reef, rank after rank passing in long perspective out of sight. Below in the clear water, the links of the cable lay as sharp on the sand as if the cutter had floated in the air. Down overboard a blue starfish (*Linckia*) slept, here a prickly urchin, there a madrepore.

Regardless of scenery the party munched their damper, and asked what was the scarlet tree ashore, and should the cook be keel-hauled for opening plum jam instead of marmalade.

"Get the dingy up and look alive!" quoth the captain. Into her tumbled all hands, Mr. Conglomerate, the captain, the cook, the jib-sheet hand, and Mr. Conchologist. Safe in the locker are stowed all the nice town clothes. A soft felt hat, flannel shirt, tweed pants, belt and sheath-knife, and big laborer's boots with thick socks rig out a man for the reef. Mem. tie your boots with string instead of

SHELLS OF JAPAN.

Explorations in Japan and the Loochoo Islands this year have brought to light many interesting and new Land Shells—fine HELICES, LAND OPERCULATES and CLAUSILIÆ—which are offered at very moderate prices. A series of Japanese Helices should be in every collection for comparison with related West American species. Also fine Marine Shells.

REFERENCE: The Editors of this Journal.   **Y. HIRASE, Kyoto, Japan.**

laces for wading. Our tools are a geological hammer, a four pound sledge, a crowbar, all the cans and buckets available, a pocket-full of corked tubes and a pocket lens fastened to the waist with a keychain, such as bank clerks use. A big shooting bag is a handy thing to sling over the shoulder. A design for reef collecting which we never put in practice was a belt like a soldier's cartridge belt to hold tubes instead of cartridges. Not only could molluscs, worms, etc., be packed apart, but such useful things as alcohol, formal, or picric acid would be at hand in small quantities. In practice we filled, say the left trouser pocket, with empty tubes. When a specimen is found it is important not to lose sight of it, and one hand may be engaged holding the rock. With the free hand a tube is taken, the cork pulled out with the teeth, the specimen bottled, the tube filled with sea water and stowed in the right hand pocket.

Now we pull in among the corals and jump overboard. "But what is the thing like a barrel stranded yonder?" "That, Mr. Conglomerate, is a *Tridacna gigas*." A real, live, giant clam, with jaws gaping like a crocodile, lying high and dry and loose upon the reef. Between the jaws are living jewels of green and gold, thick strewn on living velvet. With a convulsive jerk the shell half closes and gaps again. "I've seen plenty bigger nor this; do you want him, Mr. Conchologist?" asked the jib-sheet hand. "Yes, take that." So he drove the end of a board hard down in the centre of the gape. That disabled the monster. The cook plunged in a butcher's knife, dexterously peeled back the gorgeous mantle, slipped off the huge adductor muscle and unceremoniously threw out the carcass, bigger than a leg of mutton, on the sand.

The conchologist who ordered the execution feels, well, just a qualm of remorse, as the men hoist the shell to the boat. Anyhow he never collected a bigger shell. And then for an instant, the sunshine and the sea were swept away and the magic of memory flashed out a picture of distant lands and days; faint incense, cold and gloom, past rows of marble pillars and stained glass windows, to a small conchologist gaping with amazement at his first *Tridacna*, the the holy-water basins in St. Sulpice.

Again the ripple on the water, the sunshine and the sea. All about the giant lay lesser clams, *Hippopus*. "How do you tell one from the other, captain?" "Why, look at the meat," says he. And sure enough, we saw that *Hippopus* lacked the jewelled eyes of his great brother. Queer topsy turvy molluscs these, lying hinge down, gape up. And yet they have turned round in the shell and live heart up, foot down, like other bivalves.

"Here's something for you!" Wading across, the conchologist found four men standing at a respectful distance round one small *Octopus*. Boldly he grappled with the fearsome beast, twining and untwining the long sticky tendrils. The jib-sheet hand muttered something to the cook, and both exploded with laughter. I fear that irreverent young rascal had remarked how like the Octopus, all legs and arms, was to Mr. Conchologist himself. Now the slippery thing is gathered up and slid into the bucket. When packing time came, however, no Octopus could be found; evidently it climbed out when the gaoler's back was turned.

Another big coral block, over with it. A scuttle of little crabs, as they clatter down small holes, a shrinking of things soft, a twisting and a writhing of things neither hard nor soft; among them is a particularly energetic bunch. Left alone it unwinds into a huge Brittle star; casting a couple of cables into the water beneath, the Brittle star lowers itself along itself to the sea, pulls after itself the cords which are itself, and tucks itself, body and ropes and running rigging, comfortably into a crack in the coral.

This is a land of big things. Here is a huge sea anemone, bigger than a dinner plate, *Discosoma*, with all its tentacles spread abroad. A gorgeous little fish, crimson with a white bar, has made friends with the anemone and at the least fright swims to its capacious bosom and nestles safe among the poisonous tentacles.

And here is the Chinese dainty, the bêche de mer, a dozen different kinds of them. The commonest Holothurian is a long, black, snake-like species. When feeding they sweep all around with their branching tentacles, grasping a miscellaneous catch of foraminifera, shells and sea-weed, and thrust the mass down their throats. Another bêche de mer has earned the name of cotton-fish, because when handled it voids a mass of white glutinous threads, troublesome to clean from hands and clothes.

"Pass the crow-bar and up-end this block. A heave, my hearties, and up she goes!" " A mutton-fish," says the jib-sheet hand, and grabs it. *Haliotis asinina;* now we always did think that narrow shell could not contain the body, and here it is like *Scutus* or *Umbrella* or *Lucapina*, only a shield upon the back. "And those, toenails you call them, not a bad name either. I call this *Acanthochites* and that *Cryptoplax*." As the conchologist crooks his finger round a live *Stomatella mariei*, the creature falls asunder. Like a Gecko, he would ransom his body with his tail. *Gena* does the same in Sydney harbor; *Harpa* is said to know the trick.

"What are you doing?" asks Mr. Conglomerate, strolling up. "Smashing up *Heliopora*? What a shame!" And so it is, to wreck the beautiful blue branches, but we seek all manner of queer things hidden in corners, and *Gephyrean* worms and *Lithophaga* burrowing inside. A few odd urchins and star-fish are overhauled for *Stylifer* without success.

And see under the water what looks to the eye like a mass of white down, but to the finger feels hard. We read the riddle with a chop of the crow-bar; the fragment shows tiny crimson rods packed together, and partitioned off into floors and ceilings. It is a lump of *Tubipora musica* in full bloom.

Out in the broad daylight lie the Strombs, they love the little sandy pools among the rocks. The sociable big *Pterocera* lie around in scores, not too proud to foregather with their humble cousin *S. luhuanus*. There is nothing shy about a Stromb, it vigorously resents being picked up, and kicks like a *Nassa*, lashing out water with its operculum, and thrusting out its beautiful big, green eyes, on their long stalks, it stares boldly at its captor.

But now the tide is rising fast. Soaked and tired and hungry we must leave our hunting ground. Though bags and buckets and tubes are crammed, yet we have scarcely tasted of the riches of the reef. All to-morrow, next week, and next month we might collect without exhausting it.

Back to the cutter we row over fields of deeper corals, scarlet gorgonias, parrot fish glancing blue and gold and green, *Monacanthus* in armour of black and red, over madrepores ten feet across, like tables spread with dainty lace and edged with violets.

Then we climb aboard and snatching a hasty meal as we work, face the only tiresome labor of the day, the labelling, sorting and packing of our catch.

At last the jars and kegs are screwed down. Let us pass the pannikin along for rum, light a pipe, stretch luxuriously on the hatch and lazily watch the ghostly gleam of the zodiacal light fading in the west.

ON A TANGANYIKA BEACH

BY CHARLES HEDLEY

It is not the sea, one told oneself; but the other self insisted that it was, pointing to the level horizon where sky and water meet, to the waves rippling with the familiar song upon the sand, to the wide beach overgrown with sea-convolvulus and even to the sea gulls flying overhead. A glimpse of reeds and papyrus growing off shore is a help to appreciate that this is indeed a lake.

Not being the sea, to what can it be compared? The lake front at Chicago is recalled by the beach, the waves and the horizon. But beyond the Tanganyika beach rise range beyond range till the eye rests on an eight thousand foot mountain in the blue and hazy distance. Happily does Burton's phrase of "the lovliest glimpses of the Mediterranean" express it. But against the Mediterranean background is set a foreground which might belong to New Guinea or Tahiti, a group of blacks at play upon the shore, hills sloping steeply to the sea, green with banana plantations, dotted with brown thatched huts and flagged with palms.

In clear weather, for now is the rainy season, the hills may be seen across the lake which is only forty miles wide, though it stretches north and south for 450 miles. The surface is 2500 feet above the sea while the depth is incredible; four thousand feet is a published record, but a Belgian officer informs me (I know not on what authority) that it extends half as far again to two thousand metres. The former record exceeds the depth of any lake in the world except Lake Baikal in Siberia.

Lake Tanganyika has the narrowest zone of internal drainage, the few small streams entering it being incapable of carrying down much sediment. The exit is a small stream which runs west to the Congo. This lake then has a promise of im-

Collecting Abroad

mortality beyond any other lake whose destiny is either to be filled in by tributaries or breached and drained by exit streams. Yet it was once larger, for all around are inland cliffs composed of lake deposit and separated by a level terrace from the water.

Lake Tanganyika lies in the Great Rift Valley of Africa. This is a huge crack in the crust of the earth which dates back at least to the Cretaceous epoch. It runs for several thousand miles from the Jordan Valley in Syria through the Red Sea, Lakes Rudolph, Tanganyika and Nyassa to Cape Corrientes in South Africa.

Water in the lower strata of the lake is said to be salt. Since more water is received and evaporated by the lake than is spilled into the Congo basin, the salinity of the lake is probably increasing. My neighbor at the hotel table tells me that even the surface water is unwholesome and should not be drunk unless boiled. He also cautioned me to be careful when collecting along the beach because there are numerous and ferocious crocodiles which often snatch and kill the negroes. So impatient was I to see at last the wonderful shells of Tanganyika in their home, that I could scarcely wait till I had secured a room at the hotel and despatched a meal, before making for the lake side. When I reached it there appeared along the water side a long swathe of storm-tossed shells. These were, if I remember right, the *Neothauma*, large and solid gasteropods, some purple, some yellow, some ovate conical, some with a broad sutural furrow that recalled Eburna. This, which perhaps is a derivative of Vivipara, is a dominant form and continued in great abundance as far as my promenade extended. It was obvious that the molluscs lived close by within the sweep of the waves. The exit of the Lukuga River was not far off and as it seemed promising I walked across to it. Over a broad flood plain at the river bank were scattered great quantities of shells. Among them a large river mussel, *Iridina*, was conspicuous. On picking it up, there appeared a taxodont hinge, something strange to find in a river mussel. Where the shells were thickest I sat down to make a closer search. As I did so, I remembered being warned at the hotel that just here was where the biting flies might inflict the sleeping sickness on their victims and resolved not to linger long. The first thing I found was a large solid Melania-like shell, *Paramelania*, which brought the picture of it in Smith's paper at once to mind. Then I was rewarded with one after another of its associates. A Nassoid shell, *Edgaria*, was the most abundant; this seemed to exhibit endless variation of color, form and sculpture. There may have been several species in a handful that I gathered, but a hasty glance could distinguish no consistent features. A glossy white Naticoid was the next to attract attention, this and a rarer Crepiduloid species (*Spekia*) seemed the most un-lacustrine shells imaginable. Then I found a brown shell with a white zone, for all the world like a Planaxis, that on a coral beach would have appeared quite conformable. Most eccentric of all was a tiny white acicular shell (*Syrnolopsis*) like *Eulima* but with the plait of *Pyramidella*. These were quite common when the proper place was reached. Not one *Tiphobia* did I see; its absence is accountable by the deeper water that it is said to inhabit. Mixed with the thalassoid shells were a few ordinary fluviatile types, Lymnæa, Planorbis and Ampullaria. There were also a few land shells carried down by the storm water.

This hasty glimpse at the Thalassoid fauna of Tanganyika was ample reward for much wearisome travel. An impression remained of the extraordinary dissimilarity between this and any other fauna, and hence of its probable extreme antiquity. The lake presents itself as a refuge where biological conditions may have remained unchanged for a prolonged period. While the thalassoids slept in Tanganyika their contemporaries may have perished elsewhere through foreign competition and change of temperature and of other physical conditions. Elsewhere living fossils like Nautilus and Trigonia linger here and there, but in Tanganyika we seem to see not odd members but a whole association of species preserved beyond the common span.

Albertville, Tanganyika, Congo-Belge, Central Africa.
February 9, 1925.

Pacific Coast Land, Fresh-water and Marine shells for others from any portion of the globe. Send list to receive mine.

WILLIARD M. WOOD, 2817 Clay St., San Francisco, Cal.

LIST OF SHELLS COLLECTED ON FAYAL ISLANDS, AZORES; AND ON MADEIRA ISLANDS; WITH PREFATORY NOTES.

BY WILLIAM H. RUSH, M. D., U. S. NAVY.

William H. Rush was a surgeon in the U.S. Navy during the 1880's and contributed several articles to The Nautilus *about his shell-collecting activities. He shared his specimens with Dall in Washington and with Pilsbry in Philadelphia. A number of new species were named after him, including twelve marine species which he originally captured in small dredges attached to the anchor chains of war vessels. Dr. Rush collected extensively in the Atlantic, from Florida to Uruguay, and from the Lesser Antilles to the Azores.*

While serving as medical officer on board the Pennsylvania Nautical School-Ship *Saratoga* during the practice cruises of the year 1890 and the Spring of 1891, advantage was taken of the opportunities thus presented to collect any molluskan forms of animal life that by a little trouble and some searching could be found. No special outfit was provided other than the usual collecting box, with its forceps and knife, and the rake. Surface towing nets, made of the common bobinet, were used when sailing to collect the pelagic forms. Two dredges were kindly loaned by the Smithsonian Institution but no opportunity presented for their use.

During the cruises of the Summer of 1890, stops were made at Horta, Fayal Is. Azores; at Southampton, England, where no attempt at collecting was made although a few *H. (Trichia) rufescens* Pennant were taken at Netley Castle; and at Funchal, Madeira.

During the stay at Horta two trips were made to the small fresh water pond in the extinct crater in the mountains, for the purpose of procuring a supply of *Pisidium Dabneyi* De Guerne; but not a single specimen rewarded the visits. The journeys were made on the back of a small donkey, which knowing animal, as soon as the higher parts of the mountains were reached, selected the deep ruts in which to walk, often leaving the rider, unless with special attention upon his part, stranded ; and the pleasures of the journey were not materially increased by the community of fleas inhabiting the hangings and cushions of the saddle. To reach the edge of the crater it took four hours; the donkey with the guide, and his yelling companions, were then left to amuse themselves while the descent into the crater was made.

The crater is said to be seventeen hundred feet deep, and its sides are very steep so that the actual time of descent was quite small, but the amount of time, patience, wear and tear consumed in coming up was considerable.

At Funchal, Madeira, the ascent to the mountains was made in a vehicle upon runners and drawn by oxen over roads laid with very small paving stones and often in patterns of ornamental designs. Upon arrival at the desired elevation, the oxen were detached, allowed leisurely to make their way down, and the vehicle was pulled to one side of the road to await the time for making the return trip. That time having arrived the sledge is pulled into position, a man, holding the steering rope attached to the fore corner, stood upon each side, and then, with a shout, a push and a little confusion of the respiratory rhythm, away the whole affair goes amidst a great scattering of sparks, clouds of dust and a confusion of noises; but in an incredibly short time the trip was made that previously had taken the oxen a couple of hours to do.

ORANGE COWRIES.
(Cypræa Aurantium, Martyn.)
(Loyalty Isles.)

We have a few choice specimens of this rare and beautiful Cypræa, and can supply them at :

$10.00, $12.50 and $15.00 each, according to size.

If not satisfactory, can be returned, and we will refund your money.

A. L. HETTRICH & COMPANY,
638-640 Washington, St., San Francisco

Established 1873. California.

Vol. VII.   FEBRUARY, 1894.   No. 10.

NOTES ON COLLECTING SHELLS IN JAMAICA.

BY CHAS. T. SIMPSON.

John Brooks Henderson, Jr. was a brilliant and wealthy diplomat and author who became interested in mollusks because of Charles T. Simpson, a member of the staff of the U.S. National Museum in the late 1890's. Henderson graduated from Harvard College at the age of 13 and received his law degree from George Washington University at 16. His father was a U.S. Senator who had amassed a fortune in silver. Henderson was for a few years a private secretary to John W. Foster who was Secretary of State under President Harrison. His travels took him to Japan and South America. From the age of 30 until his premature death at 53, he was a Regent of the Smithsonian Institution. He contributed 36 articles to The Nautilus, *produced a splendid monograph on the Scaphopoda of Eastern America, and wrote a charming book about shell hunting in Cuba, entitled "The Cruise of the* Thomas Barrera." *Henderson owned the yacht,* Eolis, *which he employed to make many hundreds of dredging stations off the southeast coast of Florida. His extensive collections are in the U.S. National Museum.*

About the first of December, Mr. John B. Henderson, Jr., of Washington, and the writer visited the island of Jamaica for the purpose, principally, of collecting land, fresh-water and marine mollusks. We called on Mr. Henry Vindryes, a veteran collector and conchologist in Kingston, inspecting his magnificent set of Jamaica Shells, and receiving from him every possible courtesy and many useful notes as to localities.

As our stay was to be limited to some three weeks, we were anxious to begin work at once, to actually put our hands on some of the land snails in their homes. We hired a cab with a good natured darkey for a driver, and a miserable, little, bony horse, of uncertain color, and started for the suburbs, in the direction of Rockport with our eyes strained to catch sight of the splendid *Orthalicus undatus*, which we were told we might find on our way. The poor little horse, which wobbled about first from one side of the road to the other as if in search of snails, but probably from sheer exhaustion, was suddenly brought to a standstill without much exertion by the driver, who exclaimed as he pointed his whip to some low trees on the south of the road " Da de snail you want massa. " I think we had all observed them at the same moment, and with a shout like boys we were out of the cab and racing across the road, through a terrible hedge of wild pinguin in less time than it takes to write it. There they were, great beautiful fellows, variegated with ash color and glossy black, one, a half dozen, fifty, a hundred, in fact without limit! They clung to all kinds of trees and shrubs in the low tangled scrub, and in great numbers to the tall cylindrical Spring Cereus; in almost every case glued by an epiphragm so solid that it was well nigh impossible to dislodge them, and invariably with the spire pointing downward.

When we came out of the woods an hour afterward we were as wet with perspiration as though we had been dipped in water, and covered with every description of sticking burrs; our flesh was lacerated, and our hands dirty and bleeding, for everything in the scrub bears villainous thorns. On the debtor side we had ruined two suits of clothes, and to our credit could be placed over five hundred superb living *Orthalicus*. We had learned a lesson, too, worth remembering, viz, never wear anything decent when collecting in the tropics.

During our stay we drove around the entire island, visiting every parish. Owing to the worthlessness of our team, the illness of the driver, and the almost incessant rains we encountered on the north side, our opportunities for collecting were greatly diminished.

It was only when we stopped over a day or so at the towns that we were able to get any great amount of material. Strangely enough we found almost no marine species whatever. Occasionally on the rocky beaches we obtained *Neritina virginea*, a few *Littorinas*, *Tectarius* and *Neritinas*, but for miles, in fact along whole parishes, though the road ran near to the sea, and we watched closely, not even a valve was seen.

The lack of marine forms was made up in the abundance of the land snails, and in some cases the fresh water species. In a branch of the delta of Roaring River, under a great breadfruit tree, H. picked up a dead *Hemisinus lineolatus*. Then I looked on the rocky and sandy bottom and found it alive by handfulls, and we met with it in quantities elsewhere.

We kept an eager watch for the great white *Helix aspera*. My friend picked up a single dead specimen on the road near Falmouth, and this fairly turned our heads. We inquired of every darkey from that on, hearing of it often like the Ignis fatuüs, just a little way out of reach. Near Montego Bay we got a few more dead ones, and again as it was growing dark we discovered a dozen or so on the bushes and vines when we were nearing Savanna la Mar. The next day I started out early for a walk, resolved to find this snail if thorough search would do it. I tramped the whole forenoon and got only a few *Ampullarias*, and two o'olock found me tired, hungry, and thoroughly disgusted, seven miles from our hotel, and uncertain whether to push on to some low hills a mile ahead, or to give it up and go back. My resolve determined me and I went on. The first rounded knoll looked well at a little distance—one learns in a short time to distinguish good from poor localities a long way off. The elevation did not occupy more than half an acre; red clay with decomposed limestone. It was originally a dwarf scrub which had been partly cleaved a couple of weeks before. The first thing I saw was a fine dead *Helix aspera* on the ground, then others, there they lay thickly all around me, bright and fresh, with the animals nicely cleared out by tropic showers, the sun, and swarming insects. I hardly dared move for fear of stepping on them, and to calm my excitement, and assure myself that it was not all a wild conchologist's dream, I stood still and tried to count a hundred, but when I had got to twenty I saw half a dozen live ones clinging like a string of enormous white beads to a little shrub right beside me, and I quit counting and gathered them in. Then I sat down and without moving I picked up thirty fresh, cabinet specimens. About that time it just began to dawn on me that the great *Lucerna acuta* was as abundant as the *aspera*, and in no time I had my hands full of the fine, big, brown fellows. Afterward I got me eyes focussed down to seeing *Sagdas*, *Helix sinuata*, three or four *Cylindrellas* and as many *Tudoras*, and that under the leaves, and among rubbish there were quantities of small *Glandinas*, *Zonites* and *Microphysas*, that the ground when closely examined was literally bespangled with lovely little *Proserpinas*, that shone in the sun like polished opals.

To my dying day I never expect again to see such collecting unless I revisit Jamaica. Hunger, fatigue, headache, the flight of time were forgotten, and I was only warned that I must return by the fact that the sun was nearly down before I knew it, and that I had an eight mile walk and darkness before me. On a little spot no larger than a city lot, I had taken in a few hours over thirty species of land shells. As I reluctantly tore myself away I took fifteen *asperas* from a small Mango, and on the border of the clearing where some one had bent together a couple of young logwood trees, not as large as my wrist, I picked twenty-five more fully adult and one young one.

Shall I tell how in a narrow limestone gorge of the Rio Cobre near Bogwalk in the talus under a ledge some two rods long we found no less than forty-five species, all living, and nearly every specimen in perfect condition; or how at Mandeville the honey-combed rocks were crowded with lovely *Choanopomas*, rough as chestnut burrs, now H. wild with excitement and regardless of bats, centipedes, scorpions, and poisonous vines wedged himself into a dark cave whose mouth was at least two sizes too small for his body; how he stuck fast, and alone and far from help, could neither get forward or backward for awhile, how he pushed on to be rewarded by finding quantities of *Helix peracutissima* and the great purple *H. jamaicensis*, the latter clinging to each other on the roof like so many stalactites, a snail which, by the way, we had repeatedly been told was extinct! I fear it may be now.

SNAILS EATEN BY SHREWS. — The following shells were turned over to me for identification by C. M. Davis. These he collected in the runways of the short-tailed shrew (*Blarina brevicauda talpoides*). All of these specimens had at least the first two whorls missing. Apparently this shrew is one of the worst enemies of land mollusks in this particular section (Ann Arbor, Michigan) as quantities of gnawed shells are to be met with on all sides while collecting in the wooded sections. The shells have been identified as follows: *Pyramidula alternata* Say, *Omphalina cuprea* Raf., *Polygyra albolabris* Say, *P. zaleta* Binn., and *P. thyroides* Say.—W. J. CLENCH.

NOTES ON COLLECTING SHELLS IN CHINA.

BY JOHN B. HENDERSON.

When I saw Mr. Schmacker's splendid collection of Chinese mollusks in Shanghai, and looked over Père Heude's Unios at Sicaway, I was laying the foundation for a bitter disappointment when I took to the field myself. The great alluvial plains extending from Peking and the Gulf of Pechelli on the north to Shanghai and the lower Yangtsze on the south, are not particularly rich in species. My good friend, Mr. Schmacker assured me that "the hills" fairly trembled with molluscan life; but the hills were far away, the season unfavorable, so I continued my search along the muddy banks of the rivers and the slimy waters of the canals near Shanghai, with from fair to poor success, it being then too cold (January) for land shells.

The bulk of Chinese Unios that so closely resemble our Mississippi forms, live almost entirely in the upper waters of the great rivers and their tributaries that flow through the high lands of the interior provinces. In the neighborhood of Shanghai, *Unio murchisonianus* and *Anodonta woodiana* were the only naiads I met with, but these were generally abundant. Corbiculas and the two Viviparas, *chinensis* Bens. and *quadrata* Bens. are plentiful in the canals. I secured the services of one, Ah Sin, a bland and suave celestial, to collect for me. Ah Sin brought me, day by day, numbers of *Cyclina sinensis* wrapped in endless papers, that he assured me were rare and highly desirable Unios from the inaccessible Thibetan frontier. So Ah Sin proved a failure.

Upon a three days' journey in a house-boat from Tientsin to Tungchow (on the Pei-Ho River), I did not observe a single shell of any kind. From Tungchow to Peking, about 15 miles, I gathered quantities of *Vivipara*, *Unio* and *Corbicula* in the canal, and in the dried pools by the roadside many *Limnæa*. The walls of Peking swarm with *Cathaica pyrrhozona;* I even gathered a number of them in my bed-room, where they clung to the ceiling. This species has a wide distribution throughout China, as well as *Bithinia striatula* Bens. which I gathered in the canals about Tientsin.

Unfortunately, I had no opportunity to try the good marine collecting of the southern China coast, my only attempts for marines being in the immediate vicinity of Shanghai and at Che Foo and Taku on the Gulf of Pechelli in the north. The fauna of this region is not particularly interesting, consisting only of a limited number of boreal forms, some of which can be traced along the Aleutians to Alaska and down as far as Vancouver.

These northern waters of the Yellow Sea on the Chinese side and the Gulf of Pechelli are not conducive to molluscan life on account of the immense quantities of mud poured in by the Yangtsze and the Hwang Ho Rivers. The amount of this deposit is almost incredible. The shore line from the mouth of the Yangtsze, several hundred miles north, is a great mud bank that is rapidly extending out and filling up the shallow sea. The few Pelecypods that rejoice in such surroundings must keep awake to avoid being "snowed under." They are exceedingly difficult to obtain, and especially when the icy winter winds blow as they do in that inhospitable region. From such stations I secured an *Arca*, a *Solecurtus*, *Cyclina sinensis* and a *Solen*.

At Che Foo, where the shore is more bold and rocky, a few *Monodonta labio* and *Littorina sitchana* rewarded a diligent search at low tide.

I made a desperate attempt to take advantage of the excellent collecting in the island of Formosa, but the circumstances of my visit to that most beautiful spot was such that I found it dangerous to venture out. Some natives, however, brought me quantities of beach worn shells, out of the lot of which I selected a few fairly good specimens of *Chlorostoma argyrostomum*, *Patella testudinaria*, two *Haliotis*, and *Cyprœa isabella*.

The collector in China must be of a patient and amiable disposition to endure the throng of gaping fools that follow and ply him with a thousand questions. The quick tempered man is sure to get into trouble and get no shells.

NOTES OF A CONCHOLOGIST IN JAPAN.

BY JOHN B. HENDERSON, WASHINGTON, D. C.

It was my good fortune to accompany the Hon. John W. Foster on his diplomatic mission to Japan and China last winter. Official duties and the disabilities placed upon me by a suspicious military guard prevented me from doing very much collecting, although I eagerly seized such few opportunities that came my way to gather in the tempting array of mollusks that generally seemed near at hand.

There can be no more delightful country in the world to collect in than Japan. The natives are always pleasant and courteous and often show a disposition to assist. Even my solemn escort at times so far forgot pride and dignity as to remove their swords and wade in the muddy rice fields to capture the "dobukai." Land and freshwater shells are abundant almost anywhere. The mountain sides are especially rich in that variable Helicid group of *Eulota* (*Euhadra*) *luhuana*, its many forms and varieties. *Clausilias* cluster together in old stumps, and the rice fields fairly swarm with *Corbicula*, *Vivipara* and *Melania*. One rainy day at Nikko, a coolie brought me a branch of mulberry, upon which seventeen fine specimens of *Euhadra brandti* were crawling.

The only marine collecting I could do was at Shimonoseki, and in the neighborhood of Nagasaki. Both these localities are delightfully rich in marine forms, especially the latter point, where a larger number of the true Indo-Pacific species occurs. At Nagasaki crowds of fisherwives and their children go out every day at low tide and gather *Tapes philippinarum* Rv. for the markets. These are found in great abundance on the pebbly beaches of the bay, an inch or so under the surface. At a little fishing village called Mogi, on the Gulf of Simibara, where a small stream meets the sea, I spent two days in a conchological paradise. I shall not soon forget the thrill at my first sight of those splendid Indo-Pacific species alive and moving along, that I had only seen before in collections and figured in books. Here at low tide an exposed stretch of rocky reef was covered with *Monodonta labio* L., *Purpura tumulosa*, *Chlorostoma lischkei* Pils., *Chl. turbinatum* Ad., *Chl. rustica* Gmel., *Turbo coronatus* Gmel., and an occasional *Turbo cornutus japonica* Rve. In the crevices of the larger boulders, hidden from the light, *Euthria ferrea* Reeve clings to the rough surfaces of the granite, and *Litorina sitchana* Phil. must be sacrificed at every step. In the little pools of clear water left by the receding tide, myriads of *Umbonium* glisten in the sunlight like gems, along with the more dingy *Potamides*. Under the stones are hidden all manner of nice things—the usual *Tapes* and a quantity of small species; occasionally a pretty

Calliostoma consors Lisch., and now and then a fine, large, spiny Murex (*M. axicornis!*)

A sand- and mud-bank at the mouth of the little river is most interesting. Among a wealth of species and a profusion of specimens I stood dazed and excited. *Fusus* (two large species), with their brilliant scarlet-red bodies made furrows in the soft sand, and *Siphonalia kellettii* seemed quite as abundant. The large, fine *Polinices ampla* and *Eburna japonica* Sowb. thrive in the half mud half sand. Scattered along the shore and washed in from the deeper waters of the bay I found good specimens of *Hemifusus, Rapana bezoar, Ranella lampas, Triton tritonis, Fascolaria trapezium, Siphonalia cassidariaeformis* and *longirostris, Cassis pyrum* Lam., and *Astralium modestum* Rve.

Among a number of bivalves I remember, in particular, *Soletellina boeddinghausi* Lisch., *Tellina praetexta* Marts., *Arca subcrenata* Lisch., *Cytherea lusoria, Caecella chinensis* Desh., *Dosinea japonica, Mactra veneriformis Pecten japonicus* and *laqueatus.*

While collecting at this charming spot, I was assisted by a swarm of naked children, who vied with each other in finding specimens, and whose little, black, oblique eyes could almost see around corners.

The market places usually offer a number of the more common species, among which one can often pick out rarer and more desirable forms. *Eburna japonica* and a large *Cardium* seemed the most favored as articles of diet.

JUST PUBLISHED

NEW CATALOGUE
OF THE
Marine Mollusks of Japan.
WITH DESCRIPTIONS AND ILLUSTRATIONS OF NEW SPECIES AND NOTES ON OTHERS

COLLECTED BY FREDERICK STEARNS.
BY
HENRY A. PILSBRY.

204 pages, 11 plates. 8vo. Paper, $1.00. Cloth, $1.50.
Postage, 12 cts. extra.

PUBLISHED AT DETROIT, MICH., U. S. A., BY
FREDERICK STEARNS.

RAMBLES OF A MIDSHIPMAN

BY P. S. REMINGTON, JR.

In 1918 I had the good fortune to receive a senatorial appointment to the U. S. Naval Academy at Annapolis, and entered it with visions of opportunities to collect in foreign stations on my cruises. I went on my first cruise the following summer, in June 1919, and began to realize these visions.

Our first port of call was St. Thomas, Virgin Islands, and we were one and all glad to see the rocky shores of these islands rising sheer out of the water, after a week at sea. It was my first experience with the West Indies and I was seeing everything through a many-colored glass. To heighten the tropical aspect, we had no sooner dropped anchor off the harbor entrance than our ships were surrounded with bumboats loaded with fruit, corals, sea fans, huge *Strombus gigas*, and other things. The negroes dove expertly for coins, treading water and calling, "You heave, I dive, chief!"

I was all on fire to get ashore, and had the opportunity next day. We went in and tied up to the dock to coal, all hands donning khaki and going on liberty. With a friend, whom I had managed to interest in collecting (to his subsequent sorrow), I set off across the hills. The heat was intense and we were anxious to reach the shore for other reasons than conchological. On the way up a large hill I turned over some fallen boards near an old shanty, and was surprised to find hundreds of live *Bulimulus exilis*, in company with a few *Subulina octona*. I picked a boxful and we then moved on. That was all the land collecting I was able to do in St. Thomas, for reasons which will appear shortly.

We came out at last on a smooth beach and hastened to strip and get in for a swim. While my companion was sporting in the waves, I was scurrying over the rocks picking up

Neritas, Thais, Littorina, Planaxis, Fissurella, large *Livona pica*, and many Chitons. Compared to the drab and meager fauna of New England which I was familiar with, this was a riot of wealth.

Alas! I reckoned not on two things. The first was the West Indian Echinus. I had the misfortune to stumble through a bed of these infernal black, long-spined fellows twice, and for a moment forgot collecting. Even the voluminous vocabulary of the Navy seemed inadequate to do justice to the occasion, but the worst was yet to come.

The second thing I disregarded was the tropical sun. Those of my readers who have been in the tropics will scarcely resist smiling when I say that after our swim we stretched out on that beach *sans* raiment for *two hours* in the full glare of the sun, broken by short dips in the delightfully warm ocean! We soon repented our folly and learned to respect the power of the sun's rays.

At length we leisurely dressed, gathered up our conchological treasures, and started back to the ship, intending to visit the town also and pay our respects to the ice-cream parlor there. This last we never did, for we had not been on the road ten minutes before the punishment we had been taking from the sun's rays began to assert itself and we began to realize that we were "sunburnt"! By the time we reached the ship we felt exactly like two live broiled lobsters. In addition, we had a dizzy feeling, which we tried to ease by lying down. But we could no more lie down than fly, for we were sunburnt on every blessed inch of our bodies! After a night of misery, we presented ourselves at sick bay next morning and asked to be relieved and turned in. The doctor certainly told us what he thought of us and ordered us to be painted with picric acid and turned in. I had just forethought enough before turning in to dump my shells in a bucket and stow it in an unused bulkhead. For three mortal days we twisted in agony, getting no rest or sleep, while the skin peeled off us in slabs. Meanwhile we had left port and put to sea. But to crown my misfortunes, while I was laid up some able seaman found my bucket of shells and dumped the smelly mess overboard! I didn't mind the sunburn and the loss of liberty (though my companion did), but I did bewail the loss of my beloved shells.

Our next port was Guantanamo, Cuba, where I found all the species I had taken at St. Thomas, as nearly as I could remember, and more too. On the coral cliffs near the marine encampment I found *Tectarius muricatus* way above high water, four species of Nerita, two Littorinas, Trochus, Livona, and many others. The rocks were just paved with thousands of *Acanthopleura granulata* and *Chiton squamosus*. These last were way out where the surf was crashing and many times I had to drop my knife and hang on like a limpet while a wave broke over me and soaked me to the skin. This time, however, I took the precaution to keep my skin well covered. On a nearby beach I found *Modulus modulus* and *Cerithium atratum* in company on the eel grass, also my old friend the Echinus. Burrowing through the mud, I found several *Strombus gigas, S. bituberculatus* and *Vasum muricatum*. On the way back I noticed a white land shell plentiful on all the bushes, which I later identified as *Cepolis ovumreguli*. I also found a single specimen of *Macroceramus*, which I have not yet identified.

Of course I was much handicapped by the fact that I was in uniform, and consequently rather conspicuous, also that I had no means of properly caring for the shells on board once I had collected them. For these reasons I didn't collect as intensively as I might have, and I have often kicked myself since for it. Some day I hope to return to these interesting islands.

After leaving Guantanamo, Cuba, the squadron headed south for the Panama Canal. We passed within sight of Jamaica but did not stop, much as I should have liked to collect there. For several days we drove steadily on, manoeuvring as we went. It was a most maddening sight to me after we had made a good day's run, to see the Admiral mount the bridge and commence sending up signals for manouvres which would turn us about and start us back toward Cuba. However, schedules are inflexible things in the Navy, and we must not arrive ahead of time.

At length we awoke one morning to see the white-topped mountains of Panama coming in view over the horizon, and already we could see the indigo-blue so characteristic of the Caribbean, beginning to turn gray as we got in closer to shore. In a few more hours we were dropping anchor just inside the breakwater at Colon, and viewed the low buildings and palm-fringed shore with much interest. Alas, before we could go through the Canal we must coal ship, a job which everyone, from skipper down, cordially hates. Everyone turned out in his dirtiest clothes, officers and all, and shoveled down the shutes the never-ending piles of coal that the big cranes dropped on board. It is remarkable how much coal can be stored in a battleship. By noon next day we were through cleaning ship, and the first liberty party went up the Canal to Gatun to examine the locks and the dam. We were also taken to Coco Sola Point and shown the Atlantic defenses of the Canal. Those huge disappearing guns seemed mighty formidable to us.

When word came that we were going through the Big Ditch, all was excitement. I have been through the Canal five times since, but it still holds as much wonder and interest for me as it did the first time, and I should like to go through again. The Gatun locks, which raised us from sea level eighty-two feet to the level of Gatun Lake, are a marvel of engineering skill. It seemed strange to see a whole squadron of battleships steaming through a lake far inland, with forests and hills on either hand and pelicans flying around our boys. There is room for several more squadrons of battleships to anchor also in this great lake, made by the damming of the Chagres River. Culebra Cut, with its sheer walls towering above our fighting tops, held our interest no less than the lake had. By late afternoon we had completed our voyage from the Atlantic to the Pacific, and passed out into the latter to dock at Balboa.

On my first visit to the Canal Zone in 1919 I did not know, unfortunately, that that very efficient collector, Mr. James Zetek, was a resident of the Zone. Consequently my attempts at collecting were not well rewarded. I was also handicapped by the lack of time. We were all taken on an official party to Flamenco Island to see the Pacific defenses, and while there I strayed off to examine the breakwater for shells. What was my delight to find the rocks covered with fine large specimens of *Chlorostoma pellis-serpentis!* Further search revealed some very fine *Planaxis planicostata* and many large-sized *Nerita ornata*. The rocks were paved with large *Chiton stokesii*. The number of specimens I was able to carry was limited to what I could stuff in my pockets, as I had brought no receptacle of any kind and we were in an official party. I managed to bring away a very fair representation, however. That matter proved far less difficult than the business of cleaning them. I finally gave it up, wrapped the shells in paper, and sealed them in tin boxes.

This was all the collecting I was able to get in on my first trip, as we steamed back through the canal the next day, bound for Cuba again.

On my second visit to the Canal Zone the following year, I had the forethought to write to my old correspondent, Mr. E. P. Chace, of Los Angeles, asking if he knew of any collectors in the Canal Zone. At his suggestion I wrote to Mr. James Zetek, who kindly assured me of a warm welcome on my arrival. As we were to spend several days at Balboa, I felt certain of seeing more of the conchological treasures of Panama than I had on my first trip.

My expectations were fully realized. Once more we steamed through Gatun Lake and between the narrow sides of Culebra Cut, and docked again at the now familiar Balboa. As soon as I got shore leave, I called up Mr. Zetek and was told to come right up to his laboratory at the Public Health Department of the Ancon Hospital. I shall always remember the kindness with which he welcomed me and placed the facilities of his laboratory at my disposal. He set aside his work for the day and took me on a collecting trip to Bella Vista, where he has made so many rare finds. Not the least enjoyable part of the trip was our visit to a mangrove swamp, where we found thousands of *Cerithidea montagnei* and *C. pulchra,* together with *Littorina varia* and a fine specimen of *Linatella wiegmanni*. Deep in the mud, we could hear the *Arcas* snap their valves. Soon we came out on the beach where a wealth of species rewarded our search. Here we found three species of *Thais*, four *Anachis*, several *Cerithium, Turritella, Nerita, Litorina, Natica, Arcularia Solen, Paphia, Anomalocardia*, and many others. We filled our bags till we were weary, and the sun began to get low. Realizing that we had several hours work ahead of us to clean our catch, we hastened back to the laboratory after supper and worked till late.

A rather humorous incident, which always brings a smile when I recall it, took place on my return to the ship. It seems that the Officer of the Deck had orders to open all packages brought aboard to ascertain that no liquor was being smuggled aboard (Panama is as wet as the ocean). Consequently when I came over the gangway with my big box of shells, I was stopped and asked what the package contained. Visions of having my box opened to the vulgar gaze of the laity arose before me, and I desperately sought for means to ward off such a disaster. Finally, I summed up my courage and answered: "Sir, on my honor as a Midshipman, I have no liquor in my possession." The officer smiled and passed me on!

The next day I had the pleasure of taking dinner with Mr. Zetek and his family and enjoyed a meal cooked Panama style. Afterward I spent an enjoyable afternoon inspecting part of Mr. Zetek's collection, and was presented with a large number of his duplicates, making together with what I had collected, a very fair representation of Panamanian fauna. Mr. Zetek's remarks on the history and customs of Panama were highly interesting.

It would be a waste of time to give a list of the species collected at Panama, as Mr. Zetek's list is more complete on that point than mine would be. I refer my readers to that for more complete information.

All too soon we weighed anchor for Honolulu, a sixteen days' run, and rapidly left the jungle-clad hills of Panama behind. I was comforted, however, in my regret at leaving so congenial a country, by the knowledge that we were due to coal again at Balboa on our way back to the States, and that I would have at least one more try at the wonderful shell fauna of Panama. For the present, it was westward ho, and we all settled down to our long voyage toward the alluring isles of hula maidens and *Achatinellas!*

The second leg of our eighteen-thousand-mile cruise commenced with our departure from Balboa. Nothing untoward occurred to break the even monotony of our run to Honolulu. Very little manoeuvering was done this time, because coal is precious on such a long trip. This was said to be the longest single run ever made by a squadron of battleships, and certainly it seemed an eternity to me, for it was now the turn of our division to serve in the boiler-room and coal-bunkers, and for seventeen mortal days I stoked a boiler. I have already commented on the astonishing amount of coal a battleship can stow away, but this is far less remarkable than the way a boiler can eat it up. Never a cool place, the boiler-room usually showed a temperature of around 130° F. in the latitude of the tropics, and some poor fellow was carried out on deck every watch. Those seventeen days became an era in my life, to be looked back on with wonder and some pride.

However, all things come to an end, and at last we awoke one morning to see the extinct crater of Diamond Head showing up on the horizon. As we steamed up the bay of Honolulu, twenty-two airplanes circled over our tops, dropping aboard "leis", that Hawaiian symbol of welcome. I think that the average American citizen, who is familiar with the Hawaiian Islands chiefly through the popular music which purports to come from there, pictures Honolulu as a grove of native huts in the midst of palm trees with hula maidens dancing around to the music of ukuleles. At leart, I am inevitably asked if I saw any hula dancers on my trip to Honolulu. As a matter of fact Honolulu looks quite like any other American city with its docks, stores and business district, and native Hawaiians are far less plentiful than Japanese. Indeed they are distinctly a rarity, and although I saw one or two princesses at the naval ball, they danced a one-step and not a hula dance. I was told that there are very few real hula dancers left, and that the only good ones were not fit to be seen.

Nevertheless, there is a certain glamour about Honolulu which impresses one at once. The azure blue of the water, the rugged chains of mountains in every direction, the riot of flowers everywhere and the semi-tropical climate made a deep impression on my memory. I still consider Honolulu the most beautiful city I have ever seen, and vowed to return to it some day for a longer visit.

Before the cruise started I had had the forethought to write to Mr. Charles F. Mant of Honolulu, and the "Connecticut" had no sooner docked than he came aboard and enquired for me. By special dispensation I got permission to take dinner ashore and was treated with real Hawaiian hospitality. This last, it should be remarked, is really a substantial thing, for no midshipman could walk fifty yards along the street without being cordially asked to step into a passing automobile. Certainly I had no reason to complain, for the Mants and some Yankee friends took special pains to procure some strawberries and cook a real New England strawberry short-cake which we ate at Waikiki Beach. I shall long remember that evening.

Here, as elsewhere, I lost the opportunity to collect extensively because I was unable to secure shore leave for more than half a day. We were able to plan two trips, however, which are chalked up on my memory along with those at Panama and Cuba. The first was a hike up Mt. Tantalus along the Manoa Cliff Trail. Exchanging my uniform for khaki clothes loaned by Mr. Mant, I set out in company with that gentleman for this trail, a little beyond the city. The view on the way up was superb, and I have never seen such beautiful shades of water as the bay showed. Near shore it was a light blue with a white ring of surf, while further out it merged into a deep purple. Below us in the valleys were the neatly cultivated farms of the Japanese, some of them extended well up the hillsides. Way off to the north were other ranges of hills at which I gazed with longing eyes, for in those valleys were the Achatinellas which I had hoped to collect. They were too far away, however, to be reached in the limited time at my disposal and I had to be satisfied with an inspection of the collections of Mr. Mant and Mr. Emerson. The latter gentleman has, judging by the thousands of specimens in his collection, robbed many of the valleys of Oahu of their entire molluscan fauna. The returns on this hike with Mr. Mant were not great, but very interesting because so different from any other collecting I have ever done. In the axis of large-leaved plants we found *Auriculella diaphana* and *Auriculella castanea*, a few *Amastra turritella*, *Philonesia baldwini*, and *Tornatellides procerulus*. Under the rocks by the roadside were a few *Succinea caduca* and the introduced *Eulota similaris*. In spite of the paucity of the species taken, this hike was one of the most enjoyable I have ever made.

The second trip was a visit to Pearl City near the naval base where we got in a little marine collecting along the west locks. Here we took some fine *Nerita picea, Trochus sandwichensis, Sistrum faveolatum, Litorina scabra, Siphonaria normalis, Plecorema inaequalis, Lasmodonta sandwichensis, Mytilus crebristriatus, Cytherea lioconcha, Tellina rugosa,* and a small *chiton*. Although this represents about all the species I was able to collect in Honolulu, it should be understood that this was due to the fact that I was never able to get sufficient shore leave to make a long trip, so I blame the naval authorities and not the Hawaiian fauna. The species I collected are but a small fraction of the fauna to be found. Mr. Mant kindly made up this deficiency by generously sharing many of his duplicates with me.

A rather humorous incident took place the next day as we were going through the market. Seeing a huge pile of fine *Tapes philippinarum* for sale, I enquired of the Jap on guard what was the price of them. Pointing to one of three plates heaped high with the bivalves, he answered "twenty-five cents." When I nodded at the price, he deftly rolled a huge palm leaf into a cornucopia and poured the entire *three* plates into it. My friends laughed heartily at my discomfiture. They would have laughed still more heartily had they seen me at 2 A. M. still cleaning those shells. I kept every one, however, and got them all home.

After a week's delightful stay, we bade farewell to Honolulu and departed amid a shower of flower "leis" from the cordial islanders. I don't think I have ever been so sorry to leave a port. Its charm remains long after the memory of it has dulled.

We were now bound for Seattle, a twelve-day run. Entering the region of the northeast trade winds, we encountered cool weather and several rough days. Work in the boiler room was now a pleasure, but alas I had been shifted to a deck division, for a midshipman must become familiar with every phase of ship life on these practice cruises. So I spent the days scrubbing decks and polishing brightwork until the snow-covered Olympic range came in sight and we steamed up Puget Sound and dropped anchor in Elliot Bay.

Seattle was the one port in which I knew of no shell collector, but as I had a bona-fide brother living there, I was able to get four days' shore leave, which I made the most of, though not conchologically. Our welcome here was very warm also, for the first mail from shore brought a letter of invitation to each midshipman to dinner and a ball given in our honor. The board of commerce had called for volunteers to take care of us and each car was given a number corresponding to the one on the invitation. They even provided us with partners for the ball.

I made one attempt to collect, but the results were so discouraging that I did not try again. I had been told that there was good collecting at Fort Lawton Beach. No one seemed to know where that was, however, and after spending an afternoon in a vain hunt for it, and finding only a few *Acmaea scutum patina*, I gave up and enjoyed the more worldly pleasures that my brother offered. Later, after gaining an idea of the Puget Sound fauna through the medium of exchange, I have regretted my inactivity.

Our next port of call was San Francisco where I expected to see that very active collector, Mr. Allyn G. Smith. On my arrival, I was disappointed to find that he was leaving at once for an auto trip through northern California and Oregon, on which he hoped to discover some new things. He did, before he left, recommend a few localities where the collecting might be good and also introduced me to a bit of collecting in his own garden, for the place was overrun with *Helix aspersa*, introduced. On his advice I made a visit to Muir Woods, that home of the big trees, first ascending Mt. Tamalpais on the "crookedest railroad in the world," then down to Muir Woods by gravity car. I had only about a half-hour before the train returned, however, and though I hunted industriously through the redwood thickets, I found only one specimen of *Epiphragmophora infumata*. As this had the epiphragm I think the shells had hidden away through the dry spell. This was the extent of my collecting in San Francisco. I did not get a chance

Collecting Abroad

at the marine collecting though I wanted to try Bolinas Bay.

After coaling up once more, we headed south for San Pedro which we soon reached. Here the squadron divided, half going to San Diego, the other half to San Pedro. Fortunately, the flagship stayed at San Pedro. I say "fortunately" because Mr. E. P. Chace, with whom I had been corresponding for many years, lived at Los Angeles and I was planning to pay a visit to him. To further this plan, Mr. Chace wrote me a letter and signed it "Uncle Emory," and I had only to present this to the commander in order to get another forty-eight-hour leave of absence. I felt no compunction in doing this, considering all the fine collecting I had been deprived of through official inhibition.

That visit with "Uncle Emory" is another pleasant memory to add to an eventful cruise. Besides the pleasure of inspecting Mr. Chase's fine collection, we paid a visit to the Southwest Museum, and other points of interest. I had anticipated a taste of the collecting at San Pedro, but was disappointed to learn that the tides were not right. Mr. Chace, however, invited me to collect among his duplicates, which I gratefully did. I am not one of those, however, to whom a shell is only a shell. A specimen means much more to me if I have collected it, and I promised myself another trip to the West Coast sometime in the future.

After a most enjoyable two days at the Chaces', I returned to my ship and soon we were headed for the Panama Canal again. We picked up the rest of the squadron off San Diego and settled down for another long, hot run. Men in the boiler division, "the blackgang", told us that was the hottest run of the cruise. Fortunately I was still in a deck division. Nevertheless, I was glad when we docked at Balboa again where daily showers cool the air a little. I hastened to renew my acquaintance with Mr. Zetek and this time we made a trip to Punta Paitilla, another favorite collecting ground of his. The collecting here was the richest I have ever seen, and I made the most of it. I had hoped to get leave of absence to go to Taboga Island, where Mr. Zetek assured me the collecting is very fine, but I had exhausted my leaves and failed.

With more than regret, I bade a final farewell to Mr. Zetek and his charming family. The regret was lightened, however, by the fact that we were now going *home* and that the next port was Annapolis where a month's leave awaited us. For the last time we steamed through the Canal and regarded the now familiar sights with almost as much interest as we had the first time. We were not to arrive home without incident, though, for midway between Panama and Cuba the "Connecticut" lost both propellers and had to be towed into Guantanamo by a collier. We were transferred at once to the other ships and steamed north to Annapolis.

Although circumstances made it advisable for me to resign from the Navy two months later, I shall always be grateful for the opportunities I had to collect in various parts of the world, brief and unsatisfactory though they were. And I should like to express right here my appreciation for the fact that conchologists are such prime good fellows, no matter where one meets them. I know that my cruises were made one hundred per cent more interesting for me because of that spirit of "cameraderie" which exists among them. As one of my correspondents expressed it, "I never knew a shell collector who wasn't first-class in every respect."

SEEKING LAND SHELLS IN CUBA

BY EDWIN E. HAND

Edwin E. Hand was a Principal of a Chicago High School in the 1910's and 1920's. He was an enthusiastic amateur conchologist and used shells in his teaching of natural history. He made several field trips, one with Ferris in Arizona and one with Dan L. Emery, of St. Petersburg, to Cuba. Hand exchanged shells with many American collectors. Little is known of his personal life.

> *Don Carlos de la Torre, written about in Hand's article, was a well-known figure in Cuba for many years, not only as the leading malacologist of his time, but also as President of the University of Havana in 1921-23 and as Mayor of Havana in 1902. He was beloved by foreign malacologists, his students and relatives, but was mortally feared by his political enemies in Cuba. He was President of the American Malacological Union in 1938 when the annual meeting was held in Havana. Don Carlos had a Cuban naval cruiser sent to Key West, Florida, to pick up the 57 American malacologists and their relatives, all of whom were housed at government expense at the Royal Palm Hotel. Don Carlos published eight papers in* The Nautilus *and other works with Paul Bartsch, H. A. Pilsbry and W. J. Clench. He died on February 19, 1950, at the age of 92. His likeness is on a Cuban postage stamp, and one issue of several denominations bears several species of* Liguus *tree snails described by de la Torre.*

The writer is in doubt whether to say "*Seeking* Land Shells in Cuba" or *Finding* Dr. Carlos de la Torre. But since the two subjects are practically synonymous, let it stand *as is*.

D. L. Emery, of St. Petersburg, Florida, commanded E. E. Hand of Chicago to park his car in his garage and join him and H. E. Lowe of Long Beach, California for a trip to Cuba. All necessary arrangements being duly perfected, three enthusiastic conchologists embarked on a sturdy boat at Port Tampa on July 8th bound for the port of Havana.

Next day found us located in a hotel home and 'phoning to Dr. Carlos de la Torre. A seven-passenger Kissel Car called for us next morning and we met the King of Cuban Conchology. He is fully worthy to be placed with Arango and Poey; for his knowledge of Cuban shells is simply stupendous—unbelievable. And even better than this he is so full of "the milk of human kindness."

We had numberless examples of his brotherly love for three strangers—yes, strangers except for the shell bond. But Charles Torrey Simpson told us an incident which nicely illustrates Dr. Torre's bubbling, overflowing friendliness. Mr. Simpson said, that Dr. Torre took him by both hands (or was it shoulders?) and said, "You are Charles Torrey and I am Carlos Torre. We are brothers." And though our names are different we were treated just the same.

Our car stopped first at a tropical-fruit stand, where the host loaded up with good things too numerous to mention and treated us to our first green cocoanut milk sucked through a straw a la ice-cream soda. We also sampled mangoes and mama sapotes.

Then we drove out toward Loma de Candela and the sharp-eyed collector ordered the chauffeur to stop the car. Up and down that highway arched with century-old Ceiba and Ficus trees, we hastened knocking *Liguus* from the trunks and lower branches and seeking them in crevices where they were hiding. This was a treat never to be forgotten.

Then on again to the stone fences for *Urocoptis*, and further, at Guanabacoa a species of *Pleurodonte* planted by a collector years ago, having been brought from eastern Cuba. It may be worth while to say here that Dr. de la Torre had heard of this place, but had never visited it. So the quest was as exciting to him as to the rest of us. We drove up near the corner of the jail yard and the Doctor explained to some boys what he was looking for. The boys use these in playing games and some one was sure to know where to get them, but they knew nothing about it. Then on the diagonally opposite corner the same question was put to another group. Immediately one lad said, "Si, señor," and away he ran. In less time than it takes to tell it—he returned with a pocket full of the "caracolas"—*Pleurodonte marginella var.* Soon five grown men were scratching around a chicken yard to the bewilderment of all the hens. But we got a goodly mess and were soon homeward bound. We were about forty miles away from Havana. On the way out we stopped at a wayside inn for refreshments and returning, found at the same place an elegant course dinner, Spanish style, which the Doctor had quietly ordered as a surprise to us.

Collecting Abroad

This is a fair sample of all our days with Dr. Carlos de la Torre.

On these auto-collecting trips we ran to the coast betimes for Cerions, at Mariel, *Cerion johnsoni marielinum* galore, *C. chrysalis*, *C. salvatori*, etc., by the cigar-box full.

One day we went alone to Camoa or Jamaica and, though it was very hot and we spent a lot of our valuable time wiping sweat from our brows and "specs", we caught many *Chondropoma*, *Proserpina*, *Urocoptis*, *Eutrochatella*, *Megalomastoma*, *Helicina*, etc.

Over to Matanzas, via Hershey R. R. (yes, the chocolate man) we journeyed. Here we spent two days. One in a car furnished by Doctor Torre's brother Salvatore, collecting out by Belamar Caves and Point Sabanilla and the other with one of Dr. Torre's student assistants who took us through the Yumuri valley and Abra de Figueroa. Here *Chondropoma irradians* gave us a real thrill. It is a very beautiful species with wide flaring ruby-rayed lip. We found a few of variety *mahoganyi* —solid mahogany and well named. Many *Liguus*, on our return, rewarded our weary footsteps.

We did some marine collecting around Havana, finding *Littorina*, *Neritina*, *Tectarius*, *Fissurella*, etc. Our mentor, D. L. Emery, was poking about the Littorinas at the foot of the Prado, when he found amongst the plain brown, some speckled ones. We shall never forget the excitement six little speckled shells caused the good Dr. Torre. He looked at them carefully and exclaimed, "This is a new Littorina. This does not belong to Cuba. You are fooling me, Emery; you brought this from Tampa. A new land shell would be not surprising, but a new marine!" We all agreed it should be *emeryi* and rejoiced therein. But Doctor de la Torre was not to be caught napping and began to dig. In a few days he told us that he found, in a Porto-Rican list, undescribed and unfigured, a name that just fitted—*guttata*. After diligent searching by all of us and a student assistant, Mr. Lowe found two more.

At Guyabel we found a few of the rare Varicellas. They are carnivorous and feed on the Urocoptis, literally eating them alive.

We entrained for Guane, end of Railroad west, where Doctor Torre's letter brought us a fine reception and guides. We decided first to try for *Urocoptis elliotti* (See NAUT., May, 1908), so, a guide who knew where Bishop's Cave was located, took us there. We found no *elliotti* nor any place where we had to hang on with all fours and pick with our mouth. So we decided to do some exploring ourselves. We worked around the other side of the Sierra de Guane and became "all tuckered out." Emery and Hand were resting but the indefatigable Lowe was high on the mountain side. We two had agreed that we wouldn't climb anymore for all the *U. elliotti* in Cuba. Soon Lowe called, "Found *elliotti* and lots of 'em!" We waited not, but hastened and gathered to our heart's content. And when we got back to the hotel and looked them over we decided they were not *elliotti!* Our guide had a horse to ford us back over the river. He took three at a time and made two trips. There were five of us and one sat down flat in the mud as we unloaded. But what did we care; we were having fun!

The next day we decided to go again, alone, and see what we could find. We went where no one had ever been collecting before and Dr. Torre says we found a new species. These will be worked up later.

One day, at Mendoza in the Sierra de Paso Real and the caves in the Mogote back of the railroad station, gave us a fine lot of "*dautzenbergiana*" and some new ones. We were all tired out again and went down to the station to wait for the train. We tried to buy a ticket but they refused to sell. They tried to tell us the train was two hours late. But beyond eats, drinks and sleeps, our Spanish was a negligible quantity. But we finally got the news and hiked the four kilometers to our hotel in one hour. We took the train back to Pinar del Rio, and a wonderful bus ride of about 18 miles brought us to Vinales. This was the finest mountain scenery yet and the Mogotes and Tumbadero looked magnificent. But the hotel, to which the Doctor had directed us, had gone out of business and we had the most primitive quarters. The one-story, tile-roofed, dirt-floored dining room gave easy access to the shed in back where the caballeros put their horses, leading them in and out through the dining room. We spent two days here, had fine collecting, and wanted to stay longer but our time was going so fast we had to be moving. We returned to Havana and spent two days with the Doctor's duplicates. Dr. Torre was amused that the very thing we sought in Guane, we missed. "But," said he, "I have plenty of *elliotti* and will give you each a set." And so it was. And he gave us sets and sets, and we had to leave before he got through. But most highly prized of all he gave each of us a book entitled, "Geografia de la Isla de Cuba por Alfredo M. Aguayo y Carlos de la Torre y Huerta," and inscribed, "A guide for your future excursions and a souvenir from your friend, Dr. Carlos de la Torre." It is a book of 210 pages with many illustrations and is used in the schools of Cuba. Nearly all the places we visited are described and pictured.

In closing we give a few lasting impressions: The royal palms, dominant everywhere, the poverty of the people in the country, poor huts, dirt floors, naked children, pigs tied up to graze, drinking water hauled up from the river in a barrel and often standing in the sunshine. The kindness and hospitality of everyone, trying their best to do all in their power to guide and help you, "Americano" though you be. The highways are arched with acacias, ceibas, ficus, etc., underneath which your car glides through a shady tunnel; but from a height as you look back, you see the embowered road, a huge green serpent winding through the landscape.

It was a wonderful trip and long to be remembered, but we know three Americanos who are glad to be back in the U. S. A. And we all have his promise that Dr. Carlos de la Torre will visit us next time he comes hither. This promise and the shells we have will be an inspiration to us for years to come.

September 1, 1926.

FOR EXCHANGE.—The beautiful *Anodonta suborbiculata* Say and *corpulenta* Cp. from Thompson's Lake, Ill. Also many fine Unios from Spoon River, Ill. Fine *Helix multilineata* Say, and others. Will exchange for any species, not in my collection, land or sea.— *Dr. W. S. Strode, Bernadotte, Ill.*

A VISIT TO THE HAWAIIAN ISLANDS

BY T. D. A. COCKERELL

Theodore D. A. Cockerell, born in Norwood, England, on August 22, 1866, became one of America's most prolific scientific writers in the field of insects, mollusks and fish. For nearly 30 years he was a Professor of Zoology at the University of Colorado (1906-1934), and previous to that he was Professor of Entomology at the New Mexico Agriculture College, and Curator of the Public Museum in Kingston, Jamaica. He published over 3,000 articles on natural history, 147 of which appeared in The *Nautilus. He died at the age of 82 on Jan. 26, 1948, at Boulder, Colorado.*

Dr. C. Montague Cooke, whom Cockerell was visiting in 1924, was responsible for building up an enormous collection of land mollusks at the B. P. Bishop Museum in Honolulu. Cooke was born Dec. 20, 1874, in Honolulu, and received his doctorate from Yale in 1901, as a graduate student of A. E. Verrill. He published extensively on Polynesia pulmonate snails. Cooke was independently wealthy and financed several expeditions to various South Pacific Islands. He died at the age of 73, in Honolulu, on Oct. 29, 1948.

> "For we have seen the golden sun
> Sink in the sea,
> The moonlight on the palms, and heard
> God's melody."

It is not uncommon for modern writers, in discussing the islands of the Pacific, to begin by stating that they have little sympathy with the sentimental nonsense of previous authors, but propose to state the facts, and so forth. It is no doubt true that the expression of feeling concerning the islands has often been trifling or insincere, and at times wanting in judgment or good taste; but it should be recognized that language is barely adequate to do justice to the beauty, mystery, and romance of the subject. To the naturalist, at any rate, the splendid blue of the sea, the graceful curves of the cocoanut palms, the varied foliage of the mountain forests, are but the setting for a drama which at once stimulates and baffles his intellect. Whence came this marvellous fauna and flora? What influences have governed its evolution? Why this extraordinary multiplication of species, often with almost unbelievably restricted ranges? In spite of many learned tomes, we are still groping for light on these problems, but hopefully, convinced that their solution will furnish a key to some of the most perplexing questions confronting the biologist. With the facilities for research now established and being developed in Honolulu, we may expect to see the riddles of the Hawaiian biota attacked anew, with modern intensive methods, likely to give results of fundamental importance.

The naturalist of today is indeed fortunate, when he approaches the Hawaiian problem, in finding his path made comparatively easy by magnificent monographic works. The Fauna Hawaiiensis, the volumes of the Manual of Conchology, Jordan's lists of the fishes, give at first the impression that nearly all has been done, not only the fauna, but also the flora has received elaborate treatment. Yet in spite of all this, it is still easy to make discoveries. Last summer, while I was in Honolulu, Dr. D. S. Jordan obtained a number of fishes new to the islands (some new to science) by visiting the market each morning. Perkins, in his classic Introduction to the Fauna Hawaiiensis, stated that in his opinion only about half the endemic or truly native species of insects had been discovered and described. Dr. C. Montague Cooke tells me that he could now about double the species of *Leptachatina* described in the monograph, and he has numerous undescribed snails of other genera. Thus the present investigator is in the doubly happy position of finding plenty of important work to do, while the foundations have been securely laid by masters of the science.

In still another respect the worker of today has advantages. He finds paved roads around the islands, and facilities for getting about are vastly superior to those of former times. Dr. Perkins writes me (October, 1924): "It is hard to realize now the conditions under which I worked, from 1892 to 1897 particularly. Roads were bad, often impassible in wet weather, and I did most of my work on foot. . . I cut many paths myself single-handed through the densest wet forests, sometimes 8 or 9 miles in length, and lumped tent, guns, and all apparatus and food on my back in such places."

In the following account of my own brief explorations, I am enabled to give the names of the snails through the kindness of Dr. C. Montague Cooke, who identified nearly all my specimens. I am responsible for the slugs. The hours I spent at the Bishop Museum, looking at the wonderful collection of Hawaiian shells and hearing Dr. Cooke discuss the problems involved, are never to be forgotten. May all his learning and enthusiasm find full expression in print, where it will reach those in other lands, and of later days! He speaks at times of organizing collections and data so that posterity may find it easy to proceed. This he has done and is doing daily, but who can expect that a successor of equal ability will be found? The history of science shows that unfinished work is too often neglected, or falls into incompetent hands, as indeed is true of human activities in general. Let any one familiar with modern zoology name a dozen or twenty of the leading men now growing old, and ask himself who will step into their shoes. In too many cases one has sadly to say that no candidate is yet in sight, and none is likely to appear.

I had scarcely more than arrived in Honolulu when (July 16) Mr. O. H. Swezey, of the Sugar Planters Experiment Station, kindly offered to show me Mt. Tantalus and the upper part of the Manoa Valley. Mrs. Faus, formerly a student in our department at the University of Colorado, took us up the mountain in her car, and we circled round through the forest on an obscure trail known to Mr. Swezey, who pointed out the characteristic plants, insects and molluscs. Having been familiar with the literature of Hawaiian natural history since 1891, when I helped Dr. A. R. Wallace collect data for the second edition of "Island Life," it was with feelings readily to be imagined that I saw the *Achatinella* snails on the trees, caught the bees of the insular genus *Nesoprosopis*, observed the splendid butterfly *Vanessa tammeamea* circling through the glades, found the singular longicorn beetle *Plagithmysus pulverulentus* in its native home. The entomological results of the trip might form the subject of another article, but we are now concerned with the snails. The species collected on Mt. Tantalus were as follows:

Achatinella vulpina olivacea Reeve. This is the most widely spread form of *vulpina*, and ought to have been the type of the species. As is shown in Man. Conch., it occurs in ten or twelve valleys, the typical *vulpina* apparently only in three. My *olivacea* shells vary from green and pale yellowish to reddish-brown; usually bandless, but rarely with a peripheral black band.

A. stewartii producta Reeve. These are sinistral; remarkable for their large size. The type of *producta* is dextral, and one wonders whether the sinistral race on Tantalus (referred to in the Manual) is not an independent development. It is more robust than the dextral *producta* from the head of Manoa Valley, and looks like a different thing.

After returning from the expedition to Kauai, described below, I was much occupied with the meetings of the Pan-Pacific Food Conservation Conference, and had little time to hunt for snails on Oahu. At Kawaihapai, on the north side of the island, I found it extremely dry, and on the mountain side overlooking the sea could find only *Succinea caduca* Mighels and the introduced *Eulota similaris*. I had been brought up in England to look upon *Succinea* as almost semiaquatic, and it was surprising to find the genus in the dryest possible places. In Oahu, and particularly at Waimea on Kauai, I found hillslopes with *Opuntia*, looking exactly like places in Madeira and Porto Santo which were most prolific in endemic Helicidae. But in the Hawaiian Islands one has to go to the moist forest for the native snails, xerophytic regions proving extremely barren, with rarely anything better than some common introduced form. There are some endemic insects which inhabit xerophytic stations, but probably such areas were extremely restricted prior to the changes brought about by man. At Sacred Falls, overlooking the northeast coast, I obtained *Amastra inflata* Pfr. and *Leptachatina fragilis* Gulick. The latter appears in the Manual as a synonym of *L. gummea* Gulick, but I infer that Dr. Cooke now considers it distinct.

In the fossil bed along the face of Diamond Head, where collecting is uncomfortable on account of the burs of the grass *Cenchrus*, I found *Leptachatina oryza* Pfeiffer, and *L. subcylindracea* Cooke, the latter evidently distinct, but first published as a variety of *L. oryza*.

When I went to the islands, one of my chief ambitions was to visit Kauai, the northernmost and in some ways most interesting island. On making enquiries, I was discouraged, being told that the journey would be both expensive and difficult. Very fortunately, I learned at this point that the Trail and Mountain Club was arranging their first expedition to the island, and had no difficulty in making arrangements to go along. As the Trail and Mountain Clubs have a system of interchangeable membership, the fact that I was a member of the Colorado Club at once gave me good standing in that of Honolulu. The trip, with a company of sixteen thoroughly congenial people, men and women, was made under the leadership of Mr. R. J. Baker, a well-known photographer of Honolulu. For fifty dollars each we had five glorious days on the island, with everything found, including a large truck to take us about, and also the steamer journeys back and forth. There was even enough left over for a celebration at the Outrigger Club at Waikiki, and a ride in an outrigger canoe. Certainly the visitor who cares for out-of-door things cannot do better than get acquainted with the Trail and Mountain Club, which has a permanent office in the Young Hotel.

At a point on the eastern coast of Kauai, we left the road and took the trail leading inland, eventually reaching the Upper Anahola Valley. At different points along the trail I found snails, including *Leptachatina cylindrata* Pease, *L. striatula* Gould, *Microcystis chamissoi* Pfr., *Eulota similaris* Fér., *Opeas clavulinum* var. *hawaiiense* Sykes, and *Helicina knudseni* Pilsbry and Cooke. *Microcystis chamissoi*, except for its larger size, is extremely similar to *M. venosa* Pease, which my wife collected last summer on the Island of Rarotonga.

There is a waterfall at the head of the Anahola Valley, which the party hurried on to see, but I stopped at a sharp turn in the trail, where the vegetation was dense and conditions looked promising for snails. I was richly rewarded by the discovery of several specimens of *Nesophila capillata* (Pease), which had not been found since Pease (Am. Journ. Conch., II, 1866, p. 292) recorded it as *Helix capillata* from the "Sandwich Islands." Dr. Cooke has one of Pease's original specimens, so there is no doubt about the identity. The species is omitted from the lists in Manual of Conchology.

On the north side of Kauai are the famous Haena caves, and in this vicinity I met with *Succinea lumbalis* Gld., *Microcystis chamissoi* and *Eulota similaris*. The *Eulota* were of the bandless form (f. *unicolor* Fér.), while that from the Anahola Valley shows a brown band (f. *zonulata* Fér.). Such variation doubtless occurs throughout the range of the species.

Near the Haena caves, in the sand hills, I was able, following Dr. Cooke's directions, to find the deposit of fossil shells, containing great quantities of the variable *Carelia dolei* Ancey, which appears to be now extinct, all the known specimens being presumably from this deposit. The genus *Carelia*, with many species, is confined to Kauai and the neighboring island Niihau.

SHORE REEF HUNTING IN THE HAWAIIAN ISLANDS.

BY CHARLES F. MANT.

Being anxious to visit the shore reefs by night on the western side of this Island, Oahu, a friend and myself agreed to start on the first occasion when tides were at their lowest.

Having made all arrangements we left Honolulu at 3:00 p. m. on October 18th., and motored to Kawaihapai, a two and a half hours trip, part of the distance being over very bad roads and trails, but our little car was staunch, and we arrived safely. After a brief meal we changed into overalls, and filled our torches—large iron cylinders stuffed with a sack for wick. Then slinging our collecting bags over our shoulders we started off along the railroad track which here follows the shoreline for some miles. The scenery was very wild, the mountains coming down almost to the shore on the one side, whilst on the other the reef-lined shore stretched as far as one could see bordered with a white fringe of surf.

After about an hour's tramp we decended to the shore, and lighting one of the torches commenced our search.

At first nothing much except a few common things were found, the raised portions of reef at high-water mark being covered with thousands of *Littorina pintado*, *L. picta*, and *Nerita picea*. We proceeded a mile or so further, and then examined a rocky "flat" where the reef was full of deep holes in which brilliant little fish of many colors were swimming, whilst on the rocks were numbers of the Rough Sea Urchin (*Podophora pedifera*) the "Haukeke" of the Hawaiians who esteem this and other species as food. One had to be careful, as here and there were "blow holes" which spouted the water high into the air when a wave came in.

Presently the first "find" was made of a fine *Cypraea mauritiana*. Then a specimen of *Acanthochites viridis* was discovered on a raised coral rock. It was whilst trying to remove this shell that a big wave came in unexpectedly, knocked me over the rock, whilst my torch, collecting bag, etc. went in different directions, and a sandwich that I carried in my upper pocket was reduced to pulp! However, things were soon put right, and now we began to find the shells. In the rocky pockets were many *Cypraea caput-serpentis*, and various Cones, on the weedy rock Ricinulas, on the raised reef Chitons, Helcioniscus, Littorinas, Purpuras, nerites, etc.

The luck of the evening came to my friend who had ventured out to where the surf dashed from time to time on the large rocks, for he discovered five magnificent specimens of *Cypraea mauritiana*.

We had hoped to collect some specimens of *Cypraea reticulata* which had been found upon a former occasion, but this time we were disappointed.

It was now getting late, and the tide had turned; so we retraced our steps and returned to our headquarters, the light of the full moon making the track clearly visible.

After some supper and a change into dry things we took our blankets to the beach and slept until 5:00 a. m., being awakened by the piping of Alaska Plovers busily feeding along the shore.

The view in the early hours was very lovely, the moon still shining whilst in the distance the orange and yellow rays of the sun rising behind a bank of dark clouds with the loom of the mountains and coast beneath. For miles on either side of us stretched the shore with the blue Pacific and endless lines of snowy surf.

We started for home at 6,30, arriving at 9:00 a. m., tired and well pleased with our trip.

COLLECTING IN SOUTHERN FLORIDA, THE BAHAMAS AND CUBA

BY D. L. EMERY

Daniel Littlefield Emery, born March 28, 1861, in Wells, Maine, was a machinist who plied his trade in many states. He retired to St. Petersburg, Florida, in 1916, and for many years was one of the leading shell collectors of Florida. "Dan" was a co-founder of the St. Petersburg Shell Club, and, incidentally, an uncle of the famous Californian collector, E. P. Chace. He died at the age of 89 on April 11, 1950, at St. Petersburg, Florida. Part of Dan's collection is preserved at the University of Michigan.

During the past summer, with about two months of spare time, I decided on a trip to the West Indies. I left St. Petersburg on July tenth for Miami, where I was joined by Mr. C. C. Allen, a most ardent and thorough collector.

The east coast of Florida is a valuable field and a great many species of both land and marine shells were added to our collections. The hammocks in the vicinity of Miami are rich with the Liguus, Drymaeus and Helicina, while the so-called reefs abound with Polygyra, Urocoptis, Chondropoma, Thysanophora, Microceramus etc. The Atlantic beach furnishes us with a goodly number of Spirula and Ianthina, and the shores of the outlying keys yielded Lucina, Loripes, Iphigenia, Strombus, Modulus, Cerithium and Neritina. The shrubbery near Miami Beach fairly teemed with Cepolis, Drymaeus, Helicina and *Cerion incanum*, while on the grass we found Succinea and in the canal Perna. On the jetties at the mouth of the harbor we collected several species of Thais and Nerita, and farther up were Tectarius, Echinella, Planaxis, Littorina and Siphonaria. One side trip from Miami was by stage to Fulford, where in the sand thrown up by the dredge in a new subdivision were may varieties of marine forms in a fine state of preservation, which had been buried for many years. Among them were some of the largest and finest *Lucina jamaicensis* I have seen. The periostracum on them was almost perfect while the hinge ligament was practically gone. On the higher beach we found the same as at Miami Beach, several good Ianthina and Spirula. In the marshy land were a quantity of the *Auriculastrum pellucens*, mostly dead, and *Cyrenoidea floridana*.

After spending all the time we thought we could spare, we took passage on the S. S. Nassauvian of the Allan line for Nassau. We spent about three weeks on the island of New Providence with headquarters in Nassau. I will state here that the trip was not so much for all the species we could gather as for those from half an inch up which we could collect in quantity for our exchange lists.

Nassau proper, with its rocky and sandy beaches, its rising sunny slopes, lime rock reefs, and at the rear the dense thickets, furnishes all the varieties of collecting one could wish. On the rocky shores we found Nerita, Neritina, Leucozonia, Acmaea, Columbella, Strombus, Livona, Tectarius, Echinella, Siphonaria and many small forms not yet identified. Chitons, of two species, were most abundant. On the Sea

Grape and other trees along the shore road we gathered several varieties of Cerions, Drymaeus, Cepolis and a very few small *Oxystyla undata*. This species was one we had set our hearts on, and look as we would, for two weeks not a good live specimen came to our notice. Finally one day while out to Waterloo after Drymaeus I called my partner's attention to a very large one up in the crotch of a gumbo-limbo tree, supposing it to be inhabited by a soldier crab as all the others we had found. What was my astonishment upon dislodging him to find a beautiful live specimen. The next morning, looking over the side of the piazza over Mr. Allen's store, I had to call him up from below to see another almost as large on the side wall of the house. This proved to be alive also, making two to my credit. Of course my friend could not hide his disappointment and I could not let the opportunity for a "good jolly" get by, so had to remind him several times that it was too bad to live for two years in a place and get a little wild fellow domesticated and then have a friend come over and grab it. In some places on the island the soldier crabs are so thick that on a dull morning one can scrape them up by the bucketful. Among these we managed to find about three dozen each of fine Oxystylas.

The Drymaeus proved to be quite scarce until one slightly foggy morning on a pair of sapodilla trees in a yard of one of the natives we gathered over 150 fine specimens. We thought the find and the liberty of climbing the trees worthy of a six-pence from each of us. In the limestone walls of the quarries and some of the street cuts we found a number of perfect fossils of the *Cerion agassizii*. On the grass at Fort Charlotte *Succinea barbadensis* was very plentiful. In the cut at Union street were *Bulimulus sepulchralis, Succinea ochracina, Polygyra cereolus, microdonta* and a small Thysanophora. One day we walked across the island to the south beach, and among the palmettos were *Chondropoma revinctum* and *Cerion agrestina*. On the shore which was here both sand and rocks we found Lucina, Strombus, Asaphis, Siphonaria and Cerithiums. Here was a landing place for a conch fisherman, and in front of his cabin were heaps of thousands of the *Strombus gigas*, all with a hole on one side of the spire, where they loosen the part of the animal which is attached. These so-called conchs are a staple article of food with the natives, at the market, in the little cafes, and on all kinds of dry-goods box stands. Along the street, night and day, one can buy the little conch fritters for a penny and the smaller size for a halfpenny. As a novelty these fritters are O.K. but for a regular diet they are the limit, as I found out on my former visit when all the hotels and restaurants were closed.

Jamaica was our objective point, and after providing ourselves with a chart of the island, folding cots and mosquito bars, we waited several days for the steamer, running on a three-week schedule. At the last minute we were refused tickets on account of not having passports. A regulation wartime passport is required going from one British province to another in the same sea. We decided on a trip to Cuba, and the next sailing of the steamer for Miami saw us on the way. After another trip to Brickel's hammock and steaming out about 400 more Liguus we took the Fla. East Coast R. R. for Key West. The trip down is one certainly to be remembered, especially if one tries to study the country and ask a few questions to get a conception of the immense undertaking of constructing and maintaining a railway over these keys and the sea. It is one of the greatest engineering feats of the modern age.

At Key West the collecting is what may be termed slim. We managed to obtain a number of Cerithium, Columbella, Siphonaria, Succinea, Cerion, Polygyra and Littorina. The shore yielded about as much as could be expected with a mean tide of less than two feet.

After what I had heard of the city of Key West, I was greatly disappointed to see the sailing boats and launches which were going to ruin in the harbor, and the large number of warehouses, stores and residences with the windows and doors boarded up, cigar factories closed and the whole place with a look of decline. While here we made the acquaintance of the station master at Cudjoe's Key, and from our description of *Oxystyla floridensis* he thought they were found there. Consequently when he returned we went up with him for a day, and our search resulted in two badly broken and bleached specimens that looked as though the storms brought them. Our day was not wasted, however, as we found a goodly number of the two varieties of *Cerion incanum* which neither of us had seen before, and a Succinea not yet identified.

At Havana everything had a look of prosperity—in fact we found it the same wherever we went on the island. We made Havana our headquarters and took trips every day to some of the outside points for collecting. Santiago de los Vegas, Guanajay, Guayabal, Guanabacoa and Playa all proved good collecting ground, and at Matanzas, where we found abundant species. We visited the museum connected with the high school. This, as all other school museums I have seen, was but poorly represented in the conchological line, while the mammalia and fishes were much better. Iu the vicinity of some of the caves we had fine collecting, especially Liguus and *Cepolis bonplandi*. Everywhere we went near the shore we found Cerions in profusion. At Guayabal is a high ledge or bluff with a thick growth of hard woods and vines; here we found the best collecting. The Megalomastoma, Chondropoma and Helicina, three species, were abundant, as well as many smaller forms.

NOTES ON MARINE MOLLUSKS FROM PERU AND ECUADOR

BY A. A. OLSSON

Few personal accounts of collecting conditions along the western shores of South America have ever been published. This short description of the southern limits of the Panamanian Province appeared in The Nautilus *in 1924. Its author, Axel A. Olsson, is a distinguished paleontologist and malacologist, now living in Miami, Florida. Born in Gloversville, New York, on April 19, 1889, Dr. Olsson has an outstanding record in the production of research in marine mollusks. For years he was an oil geologist for the Sinclair and Standard Oil Companies, and now is a Life Trustee and was the* first President of the Paleontological Research Institute in Ithaca, New York.

In the course of geologic studies in Peru, opportunity was occasionally found for the collection of marine mollusks from the beaches along the northwest coast. Although these collections have not as yet been fully studied, it is believed that this preliminary account and list of the more common or otherwise interesting species may be of value. The collections begin to the south at Bayover on the south side of the Bay of Sechura and continue northward nearly to the mouth of the Tumbez river. Collections were also made at several localities on the Santa Elena peninsula of Ecuador and are included. Dall's excellent Checklist of the Peruvian Marine Mollusks has been most useful as an aid in the identification and as a guide during collecting. To this list, can be added several species and definite Peruvian locality records for several more. Species not previously known from Peruvian localities are indicated with a *.

Most of the shells contained in the collections belong to species characteristic of the Panaman faunal province, whose southern limit and border with the Peruvian province, has been variously placed as the Gulf of Guayaquil, Paita and Punta Aguja. No sharp and fast line can of course be drawn and, in the general border region of the two provinces, a certain intermingling and overlapping of the two faunas will occur. Apparently, it would seem that Punta Parinas, the most westerly point of the South-American continent, forms the most natural line that can be drawn between the two provinces, although many of the Panaman species extend beyond into the warm, protected bays of Paita and Sechura. It is at or near Punta Parinas, where the cold Peruvian or Humboldt current finally leaves the coast in a more westerly swing out into the open Pacific. The rather large list of marine mollusks collected at Negritos (near Parinas) is the result of very persistent collecting during several months of residence and does not represent the usual species found each day on the beach. The common fauna of Parinas is rather a limited one, in which the

Tivela planulata Broderip and Sowerby, is the most common and characteristic species. Among the seaweed-covered rocks exposed at low tide, are found in abundance, the *Thais biserialis* Blainville, *T. peruensis* Dall, *T. chocolata* Duclos, many Chitons, Acmaeas and an occasional *Turbo magnificus* Jonas, and *Acanthina muricata* Broderip. Going north up the coast from Parinas, we can note a progressive warming of the waters with shells becoming continually more abundant and varied. Lobitos about 15 miles north of Parinas, generally offers good collecting and at Restin and Punta Cabo Blanco still further north, mollusks have become quite abundant.

The abundance and varied character of the molluscan fauna in any locality, depends very largely upon the nature of the coast line. A long stretch of sandy beach such as that which extends south from Punta Parinas to the mouth of the Chira river, is generally devoid of much interest. Such beaches in northwestern Peru are characterized by the abundance of a few species, such as *Tivela planulata* Broderip and Sowerby, *Donax punctatostriatus* Hanley, *Olivella columellaris* Sowerby and rare shells of *Periploma planiuscula* Sowerby, *Lima pacifica* Orbigny, *Raeta undulata* Gould and *Pyrula decussata* Wood. A coast bordered by submerged rocks and ledges, as a rule offers the most favorable places for collecting. The wholly or partially submerged rocks support a large fauna mainly of gastropods; while shell-drift often very rich in the smaller and less known species, is allowed to accumulate along the sandy beaches. The coast of the Santa Elena peninsula at Salinas, is of this kind, the scene of some of the early collecting of Cuming and the type locality for many species.

Many interesting species were collected from old Indian graves which are found abundantly along certain parts of the Peruvian coast. These Indian graves are generally thickly covered with shells which for the most part, belong to the common species of the coast. Occasionally however many interesting species may be found, which are very rare on the present beaches. They include such shells as the various species of *Cancellaria*, the *Northia northiae* Gray, *Solenosteira anomala* Reeve and *Thais kiosquiformis* Duclos.

A WORD TO NEW COLLECTORS.

A gentleman who has recently interested himself in the study of shells and has commenced to form a collection, desires some suggestions as to the best course in purchasing shells. Doubtless many others are asking the same question, and a few general hints to the readers of THE NAUTILUS may not be out of place. First, last, and always, buy only from a reliable dealer. Nothing is more important to the collector whose library or experience is limited than the accuracy of the labels accompanying his specimens. A misnamed or carelessly named shell is a positive nuisance in a cabinet. Second only to the identification comes the quality of the specimen. Last of all is the price. It is far better to be sure of authentic and first-class shells than to run risks with "cheap" tradesmen. And the rarer species may invariably be obtained of the reliable dealer at much more reasonable prices, without the danger of fraud. Only the expert conchologist can safely buy from the so-called "shell-dealers." The less experienced should avoid all risk of error or fraud by making their purchases at Ward's Natural Science Establishment, Rochester, N. Y., the largest and most reliable dealers in America.

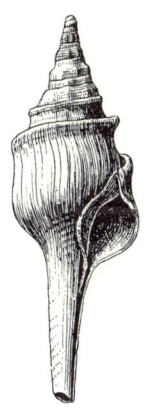

Departed Friends

The Nautilus *is a* rich source of biographical information about the conchologists who have played both minor and major roles in the history of our science in America. Over 300 obituaries and death notices have been published in the first 90 volumes. Lack of space, of course, has not permitted us to include even the most interesting of the personalities that have made our conchological history. I have included a half-dozen obituaries of various people, some famous, some only beginning amateurs, to show the concern and love expressed by the succeeding generation.

SOME CONCHOLOGICAL BEGINNINGS

BY CALVIN GOODRICH

One cannot do the necessarily repeated readings of early descriptions of fresh water mollusks in America without coming to wonder what manner of men did the original collecting, the circumstances of this collecting and, sometimes, where exactly the collections were made. Gradually, one picks up a certain amount of dependable information on the subject. A realization develops of how exceedingly restricted were the means of travel in the first fifty or sixty years of the Nineteenth Century. Customs that are now almost forgotten are revealed. Errors get themselves resolved some way. Probably more than anything else which leaves an impression is that collecting was carried on, in large part, as casually and spasmodically as the school teacher abroad gathers together her impedimenta of souvenirs.

In an incomplete list of names taken from Lea's "Observations" are those of thirteen doctors, only four of whom appear to have been collectors in the strict sense. The others pocketed a few shells as the things met their eyes, and by avenues more or less devious these shells reached Isaac Lea. This is probably why so many of Lea's types are "dead" specimens. The doctor of that day, if he lived west of the eastern seaboard or in the south, was compelled to go about on horseback. He was compelled, too, to hunt for fords, to halt at ferries, to wait with such patience as he had while floods subsided and permitted the resumption of travel. Such

---

CALIFORNIA SEA SHELLS
FOR SALE CHEAP.
SEND FOR PRICE-LIST AND SAMPLE COPY OF
"THE OLD CURIOSITY SHOP."
A Monthly Journal devoted to Science in all its Branches.
E. M HAIGHT,   -   -   -   Box 24, Riverside, Cal.

occasions were opportunities whose final results were publication at Philadelphia. Nine of the list were geologists whose duties were broad, and might range from compiling zoological catalogues to recommending sites for mill dams. To the names of two men, the "Esq." is attached. One was of the Darien district of eastern Georgia and the other of northern Alabama, and I suppose that the "esq." in the cases indicated that both were what were known as "gentlemen planters", a designation that Jefferson Davis bore at a period of his life. Three Union army officers found time amid the stresses of the Civil War to send mollusks from their stations. One was in the dust and smoke of Sherman's March to the Sea. Another barely escaped capture during an enveloping maneuver. The Rev. George White, who collected from northern Alabama to middle western Georgia, I take to have been a circuit rider, and if Bishop Elliott could not be called that, he at least journeyed about as a circuit rider did. President Joseph Estabrook had a touch of picturesqueness. He was born in New Hampshire and died in Anderson County, Tennessee. It is written of him that "he was given to elegant ruffles and fine boots, to the prodigious use of snuff, to shooting even on Fast-day and, capping all, to dreams which told him faithfully how to win $5,000 by lottery". Retiring from the presidency of East Tennessee University at Knoxville, he undertook to bore into the earth till he obtained salt. Death ended the enterprise before all his savings had been poured into the hole.

Flowing from the Cumberland Plateau in East Tennessee are two neighboring streams, Daddy's and Mammy's creeks. The old east and west road across the plateau dips steeply into a hollow of an inner fold of Walden Ridge, and in this hollow runs Mammy's Creek. I climbed afoot from Rockwood in the rain several years ago to visit the spot because the creek is the type stream of Lea's *Melania rufa*, later changed to *rufescens*. There was to be no consolation for the rain and mud. The creek was polluted with mine waste. But I am as certain as a person may be in the circumstances that this was the exact type locality. It was the only crossing made by a main road. It was such a place at which a driver would stop, water and rest his beasts and prepare against the climb up the ascent, as abrupt one way as it was the other. And here a traveler would drop from the stage to ease his legs, peer into the clear, shallow water and bring away a few mollusks which ultimately would come to Dr. Lea. Incidentally, it may be told that the name of the stream provides a measure of the literary politeness of 1841. The Mammy's Creek of the natives became Mamma's Creek in the Philosophical Society Proceedings.

The English term "watering place" that flourished for many seasons in America and finally gave way to "resort" once had a southern analogue, "retreat". Perhaps this marks an ecclesiastical influence. But in another sense the use of the word was military. For the "retreat" often was fled to as a refuge from the yellow fever of the lowlands. An early geological report of Tennessee gives several pages to describing the cool airs, the abundant brooks and freedom from disease of mountain "retreats" of that state. A lingering establishment of the kind, sprawling over a mountain top and nearly collapsing from decay, sheltered me once while typhoid ran its course, and though it could have served as a symbol of poverty it was steadfast to a tradition of hospitality and kindness. In such a place in the old days, there was naturally a great amount of leisure. Guests strolled around, and strolling they saw snail shells, maybe for the first time in their lives. The shells, by ones and twos and little packets, reached describing naturalists and became blessed binomially, some less deservingly so than others.

A northern "watering place" was Yellow Springs, Ohio. A spring gushes here from a hillside in sufficient quantity to make a good-sized stream. The place was popular with people of Cincinnati, and thither went Thomas G. Lea, brother of Isaac, who had promised he "would look after the shells of his vicinity" when he moved west. He did it so well at Yellow Springs that he sent back lots in which Isaac found two forms he announced as new. Also, Thomas roamed afield and in Buck Creek, fifteen or so miles from the springs, he took the clam we know now as *Anodontoides ferussacianus buchanensis* Lea. A footnote on Buck Creek is in Ortmann's "Monograph of the Naiades of Pennsylvania", reading, "Location unknown. Possibly near Cincinnati?" The location is given correctly on the cover of Number 8 of Conrad's "Monography of the Family Unionidae", and it seems likely that in Ortmann's copy of this work the covers had been discarded.

All sorts of possibilities develop out of the attempt to follow the course of Conrad in his journey to Claiborne and back. His intention to go south was announced upon the cover of Number 2 of Volume I of "Fossil Shells of the Tertiary Formations of North America", bearing date of December, 1832. The "New Fresh Water Shells of the United States", introduction and text, provides hazy clues to the route and the excursions in Alabama. There is further light among the descriptions in the monograph on the Unios. Other information I owe to Mr. T. H. Aldrich and to various volumes of history.

Conrad was on the James River in Virginia in March, 1833. There was then no railroad running south out of Philadelphia and the railroad between Baltimore and Washington was not opened until 1835. When John Quincy Adams went home to Massachusetts in 1825, he travelled by coach to Annapolis, crossed Chesapeake Bay by steamer, by coach again to Newcastle, Delaware, and up the bay and river to Philadelphia by steamer; time twenty-four hours. It is reasonable to suppose that Conrad reversed this journey on his own trip eight years later. After getting to the James River and to Petersburg by stage, it was possible for Conrad to go on to the Roanoke River by rail. Audubon was on this railroad in 1833, and found that passengers could not give much attention to scenery since they had to devote themselves to putting out the sparks the flew upon their clothes from the locomotive. By stage or horseback, Conrad reached Charleston. Here a railroad, in course of building to the Savannah River opposite Augusta, Georgia, could be used for seventy-two miles. This line had the first American constructed locomotive, an engine that exploded because its attendant disliked the sound of escaping steam and tied down the safety valve. Conrad, as he himself sets down, collected around Augusta. Thereafter, he had to use stage or horseback to his destination at Claiborne. He took opportunity to peek into the Oconee, Ocmulgee and Flint rivers of Georgia. The stay in Claiborne was for six months. Excursion was made to Mobile and again to the Tombeckbe River (now Tombigbee) at St. Stephens, an old capital of Alabama that has ceased to exist as a community. He visited Wilcox County, where his host in Monroe County had a plantation. By the evidence of his descriptions, Conrad was on the Black Warrior River in Green County at Erie, a village that either has disappeared or changed its name, for it is not to be found on present-day maps or in gazetteers. How he got there is not mentioned, but thanks are paid to one Dr. Robert Withers of Greene County, probably for hospitality.

Mr. Aldrich has given me Conrad's route home. It was "Claiborne on the Alabama River to Selma by boat. From Selma to Elyton (now Birmingham) by stage. Thence across the upper reaches of the Black Warrior River to Huntsville. I don't know the route from Huntsville, but presume it cut across the eastern part of the state of Tennessee to Bristol; then to Lynchburg, Washington, Baltimore and Philadelphia". About thirty miles north of Elyton was Blount's Springs (since contracted to Blount Springs). He collected near here, apparently staying for a day or two. He took naiades in the Black Warrior by prodding a sharpened stick between their gaping hinges where they lay in six to eight feet of water. Beyond Blount's Springs, Conrad collected in and on the banks of Flint River, reduced now to Flint Creek. In this same district, he assigned to the Tennessee system a tributary stream that belonged to the Black Warrior just as Lea, upon another occasion, gave to the Tennessee system a creek that belonged to the Cumberland. So far, Conrad kept to the main roads. But the "New Fresh Water Shells" shows that he swung westward as far as Tuscumbia, touching Florence, Muscle Shoals and Elk River. This digression was made easily possible by a new railroad, whose cars were drawn by horse, that joined the navigable water of the Tennessee above the shoals with that below them. It is curious that no hint of this part of the journey is given beyond Elk River "near its junction with the Tennessee". One may guess that the newer stage drivers did not linger at fords long enough to permit Conrad to stuff his pockets with shells.

In 1853, John G. Anthony, then a citizen of Cincinnati, walked through Kentucky, Tennessee and Georgia as far as Macon, "with the double purpose of renovating health, and of collecting the numerous and varied species of fluviatile shells with which our Western streams and rivers abound". Home again, he described fifty melanians, twenty-one of which were placed within the area of his journey. He "regretted his inability to give a more precise statement of habitat" than the names of states because the precautions he "had taken for keeping his collections distinct proved insufficient". Of the twenty-nine other melanians, twenty-two were assigned to Alabama, four to Ohio, one to Ohio doubtfully, one to Indiana and one to locality unknown. Unhappily, such precautions as were taken in which Mr. Anthony put trust were themselves insufficient. Walking, he could not have carried many specimens. Means of carriage for them were doubtless very inconvenient at times. The material just would get mixed. One comes to picturing him as taking up a shell and saying to himself, "Now, where did I get that?" The answer to the question was several times wrong. For example, five species that quite certainly were taken in Kentucky within the Green River drainage were written down as from Tennessee. In one case wherein Anthony broke his rule of keeping locality names confined to states and confidently assigned the species to a Tennessee creek he specified the wrong stream. Nor did confusion halt here. Mr. Anthony got the shells of his travels mixed with those from elsewhere. Three species that were credited to Alabama belonged by rights to Tennessee; one whose location was given as Indiana appears to have came from Kentucky. Tryon learned that Mr. Anthony's published localities and those of his labels did not always agree, and probably that was why he felt warranted in remarking that one of the Anthony species that was said to be from Alabama was more likely from North Carolina. But this time Anthony was right. The shell was Alabaman. Since Anthony had been in the banking business, which at least has a legend for accuracy, his mistakes may seem the harder to understand. But it must be recalled that the man was in ill health. There is a pathetic note on one of the Anthony labels, now in the Museum of Comparative Zoology: "New species det. while I was blind, by touch alone." It would be pleasant to record that this sense of touch was accurate. Unfortunately, it was not.

Two gentlemen of eastern Georgia were responsible for errors of the kind made by Anthony. Lea received from James Postell of St. Simon's Island mollusks the locality for which was given as Etowah River, one of the two big rivers forming the Coosa. Lea made four species of the lots. Three species were of the Florida "phase" which is quite distinct from that of the Coosa. The fourth seems to have come from Flint River of western Georgia, or one of its tributaries. The Flint fauna, too, is unlike the fauna of the Coosa. James Hamilton Cooper, a friend of Postell's and near enough to him to be called a neighbor, turned the Postell error the other way about. He sent Conrad six melanians as from the Savannah River. They were, in fact, from the upper Coosa River drainage.

SHELLS OF JAPAN.

Explorations in Japan and the Loochoo Islands this year have brought to light many interesting and new Land Shells—fine HELICES, LAND OPERCULATES and CLAUSILIÆ—which are offered at very moderate prices. A series of Japanese Helices should be in every collection for comparison with related West American species. Also fine Marine Shells.

REFERENCE: The Editors of this Journal. **Y. HIRASE, Kyoto, Japan.**

ANNIE M. LAW.

For much of our knowledge of the mollusk fauna of east Tennessee and western North Carolina we are indebted to two ladies, MISS ANNIE M. LAW and MRS. GEORGE ANDREWS. Before them, RUGEL had made a beginning in this beautiful but difficult mountain country. Until FERRISS and his friends began their explorations, these three enthusiasts were the only naturalists to exploit the region for land mollusks.

Miss Law[1] came from distinguished English ancestry. Her parents were John and Ann Law, of Carlisle, England. Her uncle, Richard Law, was governor of Malta. Other relatives who rendered services to the State were Chief Justice Lord Ellenborough, the Bishop of Bath and others. The Law family records were destroyed during the Civil War, so that the exact date of Miss Law's birth cannot be ascertained; but her father, John Law, came to America about 1850, Miss Law being about nine years old at that time.

"Mr. Law located some nine miles from the town of Maryville, Blount county, Tenn., a wild, mountainous country, though there was a watering-place about two miles distant where the *elite* of the South came through the summer for health and rest. Otherwise our neighbors were illiterate. There were no schools or churches, so that our parents were our teachers and companions. My father died in 1852 or '53. During his lifetime on the farm, my sister would ride into Maryville and recite Latin and algebra to Dr. Anderson (the founder of Maryville Theological Seminary) once or twice a week. Then moving into the town, she still continued these studies. She passed the examination for teaching school, and received a certificate at the age of thirteen. Being large for her age, she was given a school. She was proficient in music and gave lessons.

"Through Col. W. G. McAdoo, of Knoxville, she was introduced to Dr. James Lewis, of Mohawk, N. Y., who wished her to collect shells. She had from childhood a taste for shells, mineralogy, entomology, botany, in fact everything connected with nature. She began a correspondence with Dr. Lewis about 1868, which continued until his death. She had also a number of other correspondents in America and abroad.

"I might write a long story about some of her trips while collecting. After being in California four years, she returned to Tennessee and spent several months there. On one occasion, Dr. Lewis wished some

---

[1] The following account is from data and a biographical sketch furnished by Mrs. Fannie Law Andrews, Miss Law's younger sister.

particular shell from Bald Mountain in the Great Smoky Mountains. She procured a young man friend and two horses, and setting out from Concord, Knox county, went to the top of the Big Bald and procured the shells desired. They had to spend the night there. During the night there was a terrific thunder-storm far beneath them. She never enjoyed a trip more. Another trip she made in Monroe county was from Jalapa to Telico Plains. The distance was not great, but it was a bitter cold morning, the banks of the river covered with ice. She wore rubber boots, and wading in, got beyond her depth; but she got the shells. On this trip, I was her companion. As there was no house on the side I was on, I followed with bare feet. She suffered very much, not having dry footwear; but we soon reached a comfortable fire at a friend's, dried our things and were made comfortable. My sister never seemed to think of her own comfort when engaged in the search for shells. Many such exposed trips she made, the effects of which I think undermined her health.

"When she returned to California the second time, I was with her. We came by way of Mohawk, N. Y., visited the family of Dr. Lewis, and had the pleasure of seeing his collection. We also made a flying trip to Florida, visiting Jacksonville, St. Augustine, Palatka and Silver Springs. Wherever we were, she collected shells."

In California, Miss Law made her home at Watsonville, Santa Cruz county, with her sister, Mrs. Andrews, until her death, January 12, 1889.

Among the species discovered by her are *Gastrodonta acerra*, *Vitrinizonites latissimus*, *Polygyra chilhoweensis* and *P. lawi*.

SAD DEATH OF AN ORIENTAL BY HALIOTICIDE.—In the November *Popular Science Monthly*, Margaret Wentworth Leighton relates that while she was living in San Francisco, " A Chinamen went out on the rocks at low tide to gather some [Haliotis]. As he attempted to wrench one from its home his hand was caught between shell and rock, and so firmly held by the animal that he could not escape the rising tide and was drowned." West coast collectors should take warning. Don't fool with *Haliotis cracherodii* without having by you a crowbar or at least an ax, lest you should perish miserably like this child-like and bland Celestial.

IN MEMORIAM—DR. WESLEY NEWCOMB.

Dr. Wesley Newcomb, the last of the old school of conchologists, died at his home in Ithaca N. Y. on the 26th. of January at the advanced age of 84 years. His name belongs on the roll of honor as one of a distinguished group of American Naturalists, who made themselves illustrious by their services in the development and advancement of the study of Natural History in this country, and their contributions to scientific knowledge. Among those who may be specially regarded as Dr. Newcomb's contemporaries, the names of Gould, Binney the elder, C. B. Adams, P. P. Carpenter, Bland, Conrad, Lea, Anthony, Couthouy etc., at once occur. While Lea and Jay were among the last to pass within the folds of
" *   * the low green tent,"
before Dr. Newcomb, sad as it was, no doubt, to him, to note the loss of one old friend after another, he was happy in this, that his prolonged life, brought him in contact with many kindred spirits among the younger men and workers of the present, and he had the knowledge of their friendship and regard as a consolation.

In his early life he was more fortunate than many of his scientific friends. He had the conspicuous advantages of excellent teachers and a good education. He first attended the Academy at White Plains N. Y., and afterwards the Rennselær now the Polytechnic Institute, at that time in charge of one of the best of the earlier scientists, Professor Amos Eaton; subsequently at the Jefferson Medical College, Philadelphia, and last at the Castleton Medical College, Vermont, where he graduated most creditably. As a pupil of Professor Eaton, he was, to use his own expression "forced into the study of *shells*," in order to intelligently study *fossils*, of which otherwise he would have learned but little that was satisfactory. Living or recent shells were then termed "*Concha Marina*"—a lumping together amusingly indefinite and vague as seen in the light of to-day. To quote him further "I fancied recent shells would furnish a key to Paleontology and I expected in a few weeks of study to master the science of conchology." The result was somewhat disappointing; he found as all true students have found, whatever the path of study, that fresh vistas, eternally new, are constantly opening, and that with increased knowledge comes a wider and more distant horizon, and so like others who have the love and thirst and courage of learning, undaunted he kept right on.

His father, Simon Newcomb, of the fifth generation of the family in America, the first being Andrew who came to this country in 1635, was a physician. The son it will be noticed followed the father's profession. In 1838 he was fortunate in making a marriage that was in every way congenial; his wife a most estimable woman, his companion and friend for 54 years, survives him. After practising medicine in Albany, and a prolonged visit to the Antilles in 1846-7, in 1849 he went to California, thence to the Hawaiian islands in 1850, where he resided for five years. Here the opportunity for studying the interesting shells of the *Achatinellidae* was open to him, and he added over a hundred species to the number previously known. His exhaustive series of these beautiful forms is probably the finest extant, and the conclusions reached by him are without doubt more nearly correct, than those of other authors who have published on this rather difficult group.

In 1856 he returned to New York. In 1857 he went to Europe and part of the time had Dr. Gould for a companion. In London he had the pleasure of meeting many of the leading naturalists of the old world, Reeve, Gray, Sowerby, Adams, Hanley, Owen and others, and Deshayes, Kiener, Bernardi, Hupé and others in Paris. On his return to the United States he went to California in 1858 and established himself as a physician in Oakland, where he became well and agreeably known and soon had an ample practice. Here as elsewhere he continued his conchological studies, ever enthusiastic and ever ready to assist others as he had been from the beginning and was unto the end, all the while adding to'his collection, already magnificent, and one of the finest and best arranged in the world. His generous encouragement to collectors as well as occasional field-work himself, resulted in his adding several new species of Land, Freshwater and Marine forms to the molluscan fauna of the West Coast.

In 1867 the Newcombian collection was purchased by Mr. Cornell for the University that bears his name. Doctor Newcomb soon followed it to Ithaca and its re-arrangement and installation in the Museum of said institution, received his personal attention.

The same liberality that characterized his dealings with brother conchologists and collectors in the past, and the comprehensive system of exchanges established years before, notwithstanding the serious gaps that death had made in the list of his correspondents, continued to yield good fruit, and the collection after it had ceased to be his property, still received his fostering care, and was enriched by numerous and valuable accessions.

Dr. Newcomb was no closet naturalist wise in books yet unfamiliar with the things themselves. His erudition was inclusive and covered both. Twice he visited Europe, the chief object of his first visit being the further study of his profession; thrice he collected on the reefs in the Bay of Panama and southerly to Ecuador; also at many places in the United State of Colombia, Costa Rica, Nicaragua, Honduras and San Salvador. In 1846-7 before briefly referred to, he collected on twenty-one of the West Indian islands from Santa Cruz to Demerara, and subsequently at San Domingo, Hayti and Key West, also one winter on the Gulf coast of Florida where he made a large collection on the shores and by dredging the Sarasote Bays. In 1870 he was appointed sanitary expert to the San Domingo expedition by President Grant, the voyage being made on the U. S. S. Tennessee, and the following year, he was made one of the commissioners, to examine and report on the Sutro Tunnel, Nevada.

Dr. Newcomb was an honorary and corresponding member of many scientific societies at home and abroad. His numerous papers covering the period from 1849 to 1866, have appeared in their publications.

From the above it will be seen, how active was his life, and how enthusiastic his love of Nature. This love possessed him to the last. Of his character nothing can be said but praise. All who knew him will bear testimony to his noble sincerity and great goodness of heart, unconsciously exhibited in numberless acts of kindness, generosity and benevolence.

So closed a life well rounded with fullness of years, of good will and of generous service.
R. E. C. S.

Extremely useful to beginners and collectors for the region it covers.—Wm. H. Dall.

MOLLUSKS

OF THE

ATLANTIC COAST OF THE UNITED STATES,
SOUTH TO CAPE HATTERAS,

BY

AUSTIN C. APGAR.

THIS WORK CONTAINS A KEY TO ALL THE GENERA, A GLOSSARY OF MOLLUSCAN TERMS, DESCRIPTIONS OF ALL THE SPECIES OF SHELLS, AND OVER SIXTY ILLUSTRATIONS.

Price, Bound in Cloth, $1.00.

For Sale By AUSTIN C. APGAR, 511 East State Street, Trenton, N. J.

LORENZO E. DANIELS.

L. E. Daniels was born at Mazon, Grundy Co., Illinois, March 4th, 1852. The son of a farmer, his early life was spent on the farm, and so far as known his education was in the local schools.

While a farmer in Illinois, though a Democrat in politics, he was called from the plow in that strong Republican community to take the office of sheriff. Though modest to a fault and with none of the politician or office-holder in thought or manner, the administration was nevertheless a success. The term of office was enlivened by puzzling criminal cases, including murder, and there were also serious strikes in the coal fields; but this quiet farmer had courage, a known reputation for fair play, and was trusted by both workmen and employer. There were no complaints of violence in labor disputes during the Daniels regime.

The sheriff's rooms in the Grundy county court house at that time contained one of the best collections of Mazon creek fossils; for back in boyhood days the sheriff had become interested in those famous Upper Carboniferous beds near his home. In types, especially of insects, the collection contained many of the rarest species. They were worked up in a memoir by Dr. Handlirsch of Vienna, published by the National Museum. Mr. Daniels still owned this collection, together with the accumulations of many years of research in conchology, and the old Illinois homestead at the time of his death.

Mr. Daniels became interested in mollusks while a young man, and for many years collected assiduously, particularly in Indiana. For some years he was Assistant State Geologist of Indiana. Some of the results of his investigations during this period were published, in collaboration with Dr. W. S. Blatchley, the State Geologist, under the title "On some Mollusca known to occur in Indiana," and by Daniels alone, "A Checklist of Indiana Mollusca." Both appeared in 1903. At this time herpetology was added to his other interests, and in later trips the collection of snakes, horned toads and especially turtles claimed part of his attention.

Subsequently with Dr. Pilsbry, Junius Henderson and the writer, he was associated in field work many seasons in the

Departed Friends 245

wild places of North Carolina, Tennessee, Arizona, New Mexico, Utah and Idaho. In 1910 he joined Dr. Pilsbry and the writer in a collecting trip of several months in southern New Mexico and Arizona, and in 1914, in company with the writer, explored the Blue River region in Arizona and the Mogollon Mts., New Mexico. Many new species of *Sonorella, Ashmunella, Oreohelix* and *Holospira* were found on these excursions. In 1915 and 1916 Mr. Daniels joined forces with Prof. Junius Henderson in hunting Oreohelices in Utah and Idaho. Their results were set forth in two admirable papers, published jointly, the first exact and critical records for this fauna.

As a collector Daniels was untiring. His bag was always among the largest. He seemed to have the knack of finding unusual or abnormal shells. Some of these were illustrated by him in a special article.

Species of the molluscan genera *Sonorella, Ashmunella, Holospira, Hemphillia, Pisidium, Lymnæa*, of *Gerarus* and *Asemoblatta* (Upper Carboniferous insects), and probably other groups, have been named in his honor. His collections of land and fresh water shells, and of Mazon creek fossils are among the best.

Mr. Daniels was unmarried. Of late years he made his home with a sister, Mrs. James Foster, at La Porte and later at Rolling Prairie, Indiana. While on the farm be became interested in Masonry, often driving across the unbroken prairie a dozen miles on winter nights to attend lodge sessions at the county seat. He continued up to the thirty-third degree and the final services at La Porte were conducted by the Masonic fraternity.

In person Daniels was of the tall, strongly but loosely built Illinois type, of which Lincoln was an example. He was rather serious, but by no means lacking in humor, a good camp-fire companion. In character enterprising, interested, upright.

Seemingly in good health, nevertheless for some years he had need of a surgeon, and in October submitted to an operation at a Chicago hospital. Unforseen complications developed and he died October 23, 1918. By his death conchology has lost one of its best explorers, and his associates a loyal and loving friend.—J. H. FERRISS.

JEANETTE M. COOKE

Miss Jeannette M. Cooke died at her home on Point Loma in the city of San Diego, California, on October 21st, 1920. She was widely known among conchologists on account of the valuable material which she had accumulated from Lower California, which has gone into many of the great museums of the world and into a very large number of private collections.

She was born at Westford, Vermont, on March 10th, 1843, but went to Elyria, Ohio, when about nine years old. She came to San Diego in 1882 and opened "The World Curio Store" in 1886. This she maintained until about 1908 when she retired from business and moved to Point Loma.

Early in the history of the store she sent out a boat in charge of Captain George D. Porter and John Johnson for the purpose of collecting all sorts of marine life on the coast of Lower California. She made several changes in her boats and, about 1895, she purchased a small Chinese junk which had been built in San Diego, and which they rechristened "The World". In this boat these two men went for a more extended cruise into the Gulf of California.

Tiburon Island, the largest island in the Gulf of California, is inhabited by the notorious Seri Indians, who are the Ishmaelites of that region, their hands having been against all of their neighbors from their earliest recorded history. Miss Cooke told the writer that Capt. Porter had promised her that under no circumstances would they land on Tiburon Island. Nevertheless, about the end of October, 1896, they did land on this large island, were ambushed and killed by the Seris, and their boat was looted and burned. An investigation made by the Mexican Government at the request of our State Department elicited the fact that they had landed from a small boat and gone along the beach to collect. A band of Indians ambushed them upon their return, killing Johnson at the first fire, but Porter managed to reach the small boat on the beach and killed five of the Indians before they killed him.

After this Miss Cooke made no attempt to organize further collecting along the Lower California Coast, contenting herself with the purchase of stock and the turning over of the large accumulations of former years. She early became interested in the conchological side of her work and started many years ago to make a private collection which had reached rather large dimensions at the time of her death. She furnished the types of a large number of new species and varieties, most or all of which were described by the late Dr. R. E. C. Stearns, Dr. W. H. Dall and Dr. Paul Bartsch. Probably the most remarkable one of these was taken on Guadelupe Island off the Lower California Coast and was described by Dr. Stearns as *Uvanilla regina*. This shell seems to be a perfect *Uvanilla*, but specimens taken long after the description was written showed the operculum to be Trochoid.

In accordance with Miss Cooke's expressed desire, her private collection has become the property of the Theosophical Society and Universal Brotherhood, of which she was a member, and is held at their International Headquarters on Point Loma.

FRED BAKER.
POINT LOMA, CAL., MAY 25, 1922.

EDWARD SYLVESTER MORSE

In the death of Professor Edward S. Morse, which occurred December 20, 1925, the world has lost a great naturalist. He was born in Portland, Maine, June 18, 1838, and from childhood was a close observer of nature. At the age of 13 he had made a noteworthy collection of shells, and in 1857, at the age of 19, he published his first paper, "Description of a new species of Helix" [*asteriscus*], Proc. Boston Soc. Nat. Hist., vol. 6, p. 138. In 1859, in the same publication, he described *H. milium*. In 1859 he became one of Professor Louis Agassiz's special students at the Museum of Comparative Zoology, where he continued his studies until 1862. In 1864 he published his "Observations on the Terrestrial Pulmonifera of Maine" (Jour. Portland Soc. Nat. Hist., vol. 1). Professor Morse made the remark one day that what induced him to make a study of the small land shells was a frequent statement of Prof. Agassiz: that you will not find a very small and a large species in the same genus. In this memorable paper careful anatomical studies of these minute mollusks proved this to be true, and led him to propose seven new genera for species previously referred to the genus *Helix*. His remarkable ability as an artist, which was early exhibited, enabled him to show the radulæ, jaws and other features illustrating these genera in a clear and most instructive manner.

In 1865 Professor Morse described several new species of Pupidæ (Ann. Lyc. Nat. Nist., N. Y., vol. 3, pp. 1-6) and in 1868 he became interested in founding the "American Naturalist," in vol. 1 of which appeared an interesting illustrated paper by him on the "Land Shells of New England." About this time he made the beautiful drawings that illustrated the Binney edition of Gould's "Invertebrata of Massachusetts."

Aside from Mollusca, Professor Morse was greatly interested in the Brachiopoda and was one of the first to prove that they were not Mollusca but belonged to the class Vermes. In his papers, "Early stages of Terebratulina septentrionalis" (1871), "Embryology of Terebratulina" (1873) and "Observations on Living Brachiopoda" (1902), the plates show some of his wonderful work as an artist. Ambidextrous, he could use either hand with equal skill, and could also draw with both hands simultaneously. In lecturing, his skill with a piece of chalk was marvelous. From memory, in an instant, with a few lines, he could draw the shell or object of which he was speaking, frequently drawing both sides at the same time, and occasionally, to the delight of his audience, he drew an animal by starting at the head with one hand and at the tail with the other.

In 1871 he became professor of comparative anatomy and zoology at Bowdoin College, remaining until 1874; he also gave a series of lectures at Harvard.

In 1875 appeared a most admirable text book for beginners, "First Book of Zoology," which the present-day teachers in "nature study" would do well to pattern after. In 1876 Professor Morse was elected a member of the National Academy of Sciences and the following year received the appointment of professor of zoology at the University of Tokyo, which he filled with great success, returning to America in 1880. While in

Japan Professor Morse became greatly interested in the people, and in 1886 published "Japanese Homes and their Surroundings." His diary, kept at the time he was there, and published under the title "Day by Day in Japan," is a most interesting work and greatly appreciated by the Japanese themselves as showing the changes that have taken place in the country since that time. Professor Morse also became intensely interested in the pottery of Japan and made a remarkable collection which he described and figured in a beautiful folio catalogue. The collection is now in the Boston Museum of Fine Arts. In 1898 the Emperor of Japan conferred on him the Order of the Rising Sun, he being the first American to receive that honor.

After his return from Japan Professor Morse became director of the Peabody Museum, Salem, Mass., building up a very interesting and attractive museum, with a most artistic and instructive oriental exhibit. He was elected president of the American Association for the Advancement of Science in 1886, American Association of Museums in 1911 and Boston Society of Natural History, 1915–1919.

In the later years of his life he wrote many papers bearing on New England Mollusca which were mostly published in the Proceedings of the Boston Society of Natural History and THE NAUTILUS. His last paper—"Shell-mounds and changes in the shells composing them," appeared in *Scientific Monthly*, vol. 31, p. 429–440, Oct., 1925. In all, Professor Morse published about 40 papers pertaining to Mollusca and 10 on the Brachiopoda.

On March 14, 1910, the Boston Malacological Club was organized and Professor Morse was chosen its first president. He always took a great interest and pride in the little club and for years attended quite regularly, giving at least one paper a year, his last paper in the fall of 1924.

Professor Morse leaves a son, Mr. John G. Morse, a daughter, Mrs. Russel Robb and four grandchildren.

In closing I cannot do better than quote a paragraph taken from an article by Dr. Wm. H. Dall in *Science*, Feb. 5, 1926, which expresses so well Professor Morse's personality: "The salient characteristic of Professor Morse, apart from his devotion to science and love of the beautiful in art, was his boyish enthusiasm which captivated all who knew him. The versatility of his interests was unbounded, his love of fun overflowed at every opportunity; to meet him was to find a welcome. The world was brighter for his presence."—C. W. JOHNSON.

Vol. XII.   OCTOBER, 1898.   No. 6.

WILLIAM HENRY DE CAMP.

BY BRYANT WALKER.

Thirty-five years ago, Grand Rapids might fairly have been called the scientific center of Michigan. Through the energy and enthusiasm of a little group of men interested in natural history, the Kent Scientific Institute was organized, and a great deal of good work was accomplished in developing the fauna and flora of the western part of the State.

Prominent among the founders of the infant institution were three men, who were particularly interested in conchology, and through whose efforts the richness of the molluscan fauna of Michigan was developed with a thoroughness that has few parallels in the States west of the Allegheny Mountains.

The names of A. O. Currier, J. A. McNiell and W. H. DeCamp will always be familiar to the students of Michigan who may follow their footsteps in the field of their favorite pursuit.

By the death of Dr. DeCamp, which occurred on July 4th, the last of this little group has been called away from the activities of this life to "the unknown bourne."

Dr. DeCamp was born at Mt. Morris, Livingston County, New York, November 6, 1825. He received his medical education in the medical department of the University of New York and the Medical College of Geneva, New York, where he graduated in 1847. He at once entered upon active practice in his native State, where he remained for eight years. In 1855, compelled by failing health, he removed to Grand Rapids, Michigan, and resided there continuously until his death. From 1855 to 1857 he was engaged in the drug business, but having been ruined by the destruction of his store by fire in that year, he resumed the active practice of his profession in which he continued until overtaken by his last illness.

He made a specialty of surgery and acquired a large and successful practice. He was a member of the American Medical Association, the Michigan State Medical Society and the Grand Rapids Medical and Surgical Society, and, by the latter two, was, at different times, honored with the presidency. He was the author of a number of papers on medical and surgical subjects, which appeared in the proceedings of these societies and in different medical journals. He was also a member of the American Association for the Advancement of Science, the Academy of Natural Sciences of Philadelphia, and other scientific societies.

Upon the breaking out of the war in 1861, he was commissioned surgeon of the First Michigan Regiment of Engineers and Mechanics, and remained in the service until 1864, when he was mustered out with his regiment. During the winter of 1862-3, he was Post Medical Director at Harrodsburg, Ky., where 1,500 Confederate wounded had been left by General Bragg in his retreat from Kentucky after the battle of Perryville.

From an early day, Dr. DeCamp was an active and enthusiastic student of natural history. Geology, botany, ornithology, entomology and conchology all received his attention and contributed to the fine collection which, in course of many years collecting, was accumulated by him.

It was conchology, however, that, from the time of his removal to Grand Rapids, especially occupied his attention, and his work in this department will be his most lasting monument.

He was an assiduous collector. During his army life he took advantage of his opportunities in the south to pursue his favorite study and thereby acquired many interesting species. This material was forwarded by him to Mr. Currier, and by the latter to Dr. Isaac Lea and other eastern naturalists for determination. A somewhat hasty review of the literature has shown that the following new species were discovered by him during this period:—

Pleurocera currierianum Lea.
Pleurocera bicinctum Tryon.
Goniobasis decampii Lea.

Goniobasis louisvillensis Lea.
Goniobasis informis Lea.
Eurycœlon leaii Tryon.
Campeloma decampii W. G. Binn.
Somatogyrus currierianus Lea.
Unio depygis Con.

Two new species were added to the fauna of Michigan from his collection, viz.: *Succinea decampii* Tryon and *Vertigo morsei* Sterki. The types of *Limnæa desidiosa* var. *decampii* Streng, recently described in THE NAUTILUS, were also found by him.

In 1881, under the auspices of the Kent Scientific Institute, Dr. DeCamp published an elaborate "Catalogue of the Shell-Bearing Mollusca of Michigan." This, which is his only publication in conchology, contains a list of 221 species and 9 varieties, and was the most complete list of the State fauna published up to that time. Eliminating synonyms and doubtful forms, it gives a total of 185 species as now recognized as against 149 species cited in Currier's catalogue of 1868. It also is of particular value as containing descriptions and figures of three species named but never formally described by Currier, viz.: *Limnæa contracta*, *Physa parkeri* and *Anodonta houghtonensis*. He was an enthusiast in his scientific work, and his time and collection were always at the service of his fellow collectors.

Through his generosity the first set of his Michigan shells, upon which his catalogue was based, is a cherished part of the writer's collection, and the remainder of his shells have been deposited in the Kent Scientific Institute, where they "will be kept to benefit and instruct those who come after him."

WM. D. HARTMAN, M. D.

Dr. William Dell Hartman, whose death occurred on August 16th, at West Chester, Pennsylvania, was born in East Pikeland township, Chester county, Pa., December 24, 1817. He was the eldest son of General George Hartman. The founder of the Hartman family in Chester county was his great-great-grandfather, John Hartman, a native of Schwerin, Hesse Cassel, Germany, who came to Philadelphia in 1753.

Departed Friends

After receiving an education in the schools of the neighborhood, William Hartman attended the famous school of Jonathan Gause and the academy of Jonathan Strode. He studied medicine with Dr. Wilmer Worthington and about that time became profoundly interested in the study of botany. Dr. William Darlington, in his "Flora Cestrica," mentioned him as "a zealous and promising young botanist." He attended the University of Pennsylvania and graduated from the Medical Department in 1839, at the age of twenty-one years. After graduation he returned to West Chester and engaged in the practice of his profession. His practice soon became very extensive and was maintained until the infirmities of advanced years compelled him to relinquish it. Even then many patients visited him at his office.

Dr. Hartman devoted all the time that could be spared from his medical practice to the study of natural science. Besides botany, he studied entomology, mineralogy and conchology, and became an authority in each of these branches, especially as they related to Chester county. The latter science, however, finally engrossed Hartman's attention; and it is by his work on mollusca that he became most widely known. For over forty years he corresponded upon conchological topics with the leading authorities in the science in this country and abroad. Among his correspondents and co-laborers were Isaac Lea, LL. D., of Philadelphia; Charles Wheatley, of Phœnixville; George W. Tryon, Jr., J. G. Anthony, Dr. J. C. Cox, Andrew Garrett, W. H. Pease, E. L. Layard and many others.

Through Mr. Garrett he obtained what is perhaps one of the finest collections of Polynesian land shells in the world, especially of the genus *Partula*, of which he described 25 species. The critical study of the genus *Partula* particularly engaged his attention during several years; and his careful work and extensive writings on the group, have given him high rank as an authority thereon. His beautiful collection of *Achatinellidæ* was purchased by the Bremen Museum some time before his death. The last conchological work that Dr. Hartman undertook was a revision of the *Helicinidæ*. He gathered much interesting material, but failing health prevented the carrying out of his intentions. His collection of marine shells is rich in Polynesian species and contains many varieties.

One of the best known publications from the pen of Dr. Hartman was "Conchologia Cestrica," treating of the mollusca of Chester county, Pa. In its preparation he was associated with the late Dr.

Ezra Michener, of New Garden township, but Dr. Hartman did the larger part of the work. It is illustrated with 207 wood-cut figures of shells described in the work. This book, published in 1874, was formerly used as a text-book in many schools in Chester and Delaware counties.

Dr. Hartman was elected a corresponding member of the Academy of Natural Sciences of Philadelphia in 1853, and the same year he was elected a corresponding member of the Lyceum of Natural History of New York city. He was a member of the West Chester Historical Society and one of its organizers; also a member of the West Chester Philosophical Society, the Geological Club of Chester County, and an honorary member of the Chester County Medical Society. He was a member of the Episcopal Church.

In 1883 Dr. Hartman went abroad, visiting the British Museum, Jardin des Plantes and other museums.

Personally Dr. Hartman was of a most kindly disposition and social so far as his busy life would allow. His kindness professionally and to his scientific friends and co-workers was unlimited. He was ever ready to assist young men in their studies in the various branches of science, and many owe their first impulse to his encouragement and kindly assistance.

Dr. Hartman married Mary Jane Kabel, a daughter of John Kabel, of Jefferson county, West Virginia, on December 3, 1841. Mrs. Hartman survives her husband, as do also five children.

OBITUARY.

MRS. D. L. GARLICK expired suddenly March 16, 1894, in San Francisco. She was spending the winter in Alameda, as the guest of her sister, Mrs. Gaylord, 2116 Central Avenue, and the two ladies went out to the Cliff House and vicinity yesterday to spend the day collecting shells.

They climbed a precipitous height near Land's End station on the line of the Ferries and Cliff House Railroad, and when she reached a little station on the road, they sat down for a rest. Suddenly Mrs. Garlick fell forward and dropped on the floor. Upon trying to raise her up, Mrs. Gaylord was horrified to see that her sister was dead.

Weakness of the heart, aggravated by over exertion, is attributed as the cause of death.

JOHN FORD.

John Ford was born in Chester, Pa., November 15, 1827, and died in Philadelphia, January 10, 1910. He was the son of Lewis H. and Esther (Ogden) Ford. His father died when he was about ten, and he was then practically compelled to shift for himself. He lived for about three years with a farmer in Delaware county, Pa., and then returned to Chester and entered a store. By untiring efforts the boy acquired a substantial education. His great love of nature and of music was a powerful incentive to self-cultivation.

On September 6th, 1847, Mr. Ford married Phoebe T. Flavill, of Chester, Pa. After living for a time in Paterson, N. J., and Shipman, Ill., he returned to Philadelphia in 1861, shortly after entering the Corn Exchange National Bank, where he was continuously employed until July, 1903, when, owing to a severe illness, he was compelled to retire from active business.

Mr. Ford was afflicted for many years by an ever-increasing deafness, which caused him to avoid social gatherings, and in his later years even scientific and musical meetings. Yet his warm and generous nature found pleasure in the company of a few intimate friends who shared his tastes.

Mr. Ford was an accomplished musician, many of his compositions having much merit. He especially excelled in melody. Many of his songs written over thirty years ago are still in favor. His first published song, "Will You Love Me when I'm Old?" had a tremendous and immediate success, and was by far the most popular of his compositions, though a number of those appearing later were deemed more worthy of consideration by his musical friends. Among these were "Watching and Waiting," "Daisy and I," "Away Down South," "Sweet Rosalie," and, best of all, "My Ships are Coming Back to Me"—a piece of deep poetic feeling.

As a young man, Ford was deeply interested in geology and mineralogy. A warm friend of Conrad, together they collected cretaceous fossils from the marl beds of New Jersey. With the late Theodore D. Rand he collected the minerals of Delaware and Chester counties. In the early sixties Ford met the Rev. Dr. E. R. Beadle. This acquaintance ripened into a warm friendship, and turned his attention from mineralogy to conchology. He often said—" It was Dr. Beadle who started me in the study of conchology, and who took all my minerals in exchange for shells." George W. Tryon, Jr., was also his warm friend, and their bond of friendship was strengthened by their love of music. Another true and life-long friend who survives him is Mr. Charles Morris, of Philadelphia.

Mr. Ford was elected a member of the Academy of Natural Sciences in 1866 and from that time until his illness in 1903, took an active interest in the Academy's magnificent collection of mollusks. He was especially interested in the local species and those of the New Jersey coast, making a special exhibition collection of both, for the Academy and also for the Wagner Free Institute of Science. Aside from the local shells he was also especially interested in the *Olividæ*. His long suites selected to show specific variation are unequaled in this country.

He also brought together a very fine collection of *Cypræidæ*, which although containing none of the great rarities, is notable for its large series illustrative of variation. His entire collection is remarkable for the perfection of the specimens and the great number of representative genera and species of all the principal faunal regions. He had little interest in fresh-water shells, aside from local species, or for the small or minute land snails, though he possessed a very fine series of exotic Helices and Bulimi, numbering many forms now very rare. The marine pelecypods such as *Veneridæ*, *Cardiidæ*, Pectens, etc., of his collection are especially fine.

Mr. Ford published 29 articles on conchological subjects, besides a number of short notes, and a few articles dealing with paleontological and archaeological topics. His influence was largely personal. At the time of his greatest activity in conchology—1870 to 1895— he maintained a large correspondence, and was always ready to assist students with information from the rich library of the Academy. He was particularly helpful in naming specimens for conchologists who had no access to large collections or libraries, much of his leisure being given to this generous propaganda. Several species were named in his honor, among them *Donax fordi* Conrad, *Cerion fordi* P. & V., *Drymœus fordi* Pils. *Pleurodonte fordiana* Pils., *Phasianella fordiana* Pils., and others.

The portrait accompanying this sketch, is from his last photograph, taken when Mr. Ford was about sixty years of age.

He is survived by two sons William Henry and Albert W. Ford and two daughters Parthenia Ford and Mrs. Charles W. Johnson.

DR. JOHN CLARKSON JAY.

Dr. John Clarkson Jay, a son of Peter Augustus Jay and grandson of Chief Justice John Jay, a distinguished member of the First Continental Congress, died at his home, "Rye," at Rye, Westchester County, N. Y., on Sunday, being in the eighty-fourth year of his age. The immediate cause of his death was senile gangrene. Mr. Jay was graduated from Columbia College in 1827, and afterward took his diploma as M. D. Upon his marriage with Laura Prime, a daughter of Nathaniel Prime, a well-known banker, he left the practice of medicine and for a short time was engaged in the banking business, but in 1843 retired from both business and professional pursuits, to live at the country seat at Rye, on Long Island Sound, left to him by his father's will. This beautiful residence gave him full occupation, as it embraced upward of 400 acres of land.

Dr. Jay was well known in the scientific world as a specialist in Conchology, and his collection of shells was for many years the most noted in the United States. It was purchased several years ago by Miss Catharine Wolfe, and presented by her to the American Museum of Natural History.

Dr. Jay was for many years a trustee of Columbia College, was one of the early presidents of the old New York Club, and was one of the founders of the New York Yacht Club. He was a Republican in politics, and one of the early members of the Union League Club of this city.

Dr. Jay was also actively interested in the Lyceum of Natural History (now the New York Academy of Sciences) and was its Treasurer from 1832 to 1843. At this time he was a man of twenty-five or thirty, of light complexion, open and pleasing countenance, and somewhat nervous temperament. During his more vigorous years Dr. Jay was much interested in aquatic sports and was the owner of a famous yacht called "Coquille." The valuable addition to the treasures of the Natural History Museum purchased by Miss Wolfe is now known as the Jay Collection. The shells gathered during the expedition to Japan under command of Commodore Matthew C. Perry were submitted to Dr. Jay and he wrote the article on them that appeared in the Government Reports. Dr. Jay was the author of "Catalogue of Recent Shells," which was published here in 1835; "Descriptions of New and Rare Shells," and of later editions of his Catalogue, in which he enumerated about 11,000 well-marked varieties and about 7,000 well-established species.

YOICHIRO HIRASE

A short time ago we received word from Mr. Shintaro Hirasé, son of Mr. Y. Hirasé, of his father's death, which occurred May 25, 1925.

Few men, and certainly no one in Japan, have done more to advance the study of Mollusca than Mr. Hirasé. His enthusiasm, perseverance and sacrifice for the science of conchology is best described in the following paragraph, taken from a leaflet asking for aid and support for his Conchological Museum:

"At first I used to go myself collecting in different parts of the empire, but finding it very difficult because of a weak constitution, to adapt myself to circumstances, I decided to employ and educate two or three assistants, in spite of limited means, and send them not only to all parts of Japan, but also to many far away groups of islands, such as the Bonins, Loochoos, the Kuriles and Formosa; also to Korea and China, with the view of collecting material for study. The expenses of these explorations amounted to a considerable sum. As I pursued my studies I needed books and magazines which cost a great deal. On the other hand I tried to publish a conchological magazine and a few other books in order to announce to the public the results of my investigations and to disseminate information of newly discovered facts. In all of this work I spent one-half of my property."

In the NAUTILUS, Vol. 27, June, 1913, was published an account of the opening of The Hirasé Conchological Museum, with a picture of the building. It certainly required great enthusiasm and optimism to establish a purely conchological museum,—the only one of the kind ever attempted.

From 1907 to 1909 Hirasé published the "Conchological Magazine," Vols. I and II and four numbers of Vol. III appearing. It contains many fine illustrations. In 1914 he started a unique and interesting publication in Japanese style, "Illustrations of a Thousand Shells." Three volumes were published, containing 300 beautiful colored figures. The "Album of the Hirasé Conchological Museum" and the "Terebridae of Japan" are some of his publications.

Through his publications, correspondence and the distribution of shells Hirasé became known to conchologists all over the world. His investigations resulted in a great increase in the scientific knowledge of the Japanese fauna. Many species of mollusks have been named in his honor, and a peculiar genus of land snails, *Hirasea*, perpetuates his name.

IN MEMORIAM—ROBERT WALTON.

It is with sad hearts that we record the death of our young friend Robert Walton. While out collecting on Saturday, November 11, along the steep bank at West Conshohocken, he slipped and fell as a freight train was passing below, receiving a terrible gash on the head and having one of his legs crushed beneath the wheels, from which he died at 8 P. M., the accident occurring about noon. He was born in Halifax, England, July 17, 1875, and came to this country in the summer of 1889. He was a collector from boyhood, studying nature with that enthusiasm which only a born naturalist can. He was not content with a collection of shells alone; his was a collection of the mollusca. He studied their anatomy, working out their jaws and dentition, the darts from the *Zonites*, and the testaceous shell-plates, from the *Limaces*. He was a close observer, and by his zealous collecting he found many forms not before recorded from this section. Among his rarities were reversed specimens of *Zonites cellarius* and *Zonites ligerus*, and I remember his saying, when we met only a few days before his sad accident, that he found the reversed *Zonites cellarius* at West Conshohocken. He was to be appointed as a Jessup student at the Academy of Natural Sciences, and was looking forward, as only a young heart can, to the day when he would be studying and working there among the objects he so dearly loved. Mr. Pilsbry was looking forward with a great deal of pleasure to the time when he would have such a valuable assistant. We shall miss him with his bright and happy face and his pocket full of shells, and all tender our heartfelt sympathy to his parents and brothers.

The appended lines are by his friend, Mr. John Ford.

C. W. J.

Toll, toll the bell! his young heart beats no more;
   His eyes are dimmed, his life's short cycle run.
No more may Science yield him, as before,
   The charming favors he so fairly won.

Alas, that in the East his sun should set,
   And 'neath the shadows hide the hopes he knew!
Bright hopes, recalled to mind with keen regret
   By all who felt his power to will and do.

Though now in sorrow we must say "Farewell!"
   Sweet memories of him our hearts will hold;
While through the years that Time for us may tell
   His friendship shall be cherished as of old.

Dr. Herbert Huntington Smith, Curator at the Museum of the University of Alabama, was killed by a train on March 22. A notice of his life and work will appear later.

The following extract from a letter written to the Ed. by Dr. W. D. Hartman, will be of interest:

"I have just learned through Mr. Rossiter, of the Island of Noumea, that Mr. de Latour and his son (from whom I have received so many new shells from Aura Island, New Hebrides) have been murdered by natives; Mr. Garrett was wont to tell me of the great danger to be encountered by these collectors in these islands from the natives. When he was collecting in some of these islands he was obliged to be a walking arsenal and would never trust a native behind his back for fear of being stabbed and dragged off into the bushes and eaten.

I much regret the loss of de Latour as a collector. The last box he collected was lost in a vessel that was wrecked, and after floating about on the ocean was wafted to shore, and was found and sent to Mr. Rossiter."

Some of the shells contained in this box were figured in Dr. Hartman's last paper in the Proceedings of the Academy of Natural Sciences of Philadelphia.—ED.

ELIZABETH LETSON BRYAN, SC. D.

Elizabeth Letson Bryan died on February 28th at her home in Honolulu, of an organic heart affection after an illness of nearly eight months.

Mrs. Bryan was born April 9, 1874, at Griffin's Mills, Erie Co., New York, the only child of Augustus F. and Nellie Webb Letson. She was a direct descendant from Governor Bradford, first governor of Massachusetts, and was a member of the Mayflower Society of New York. She early became interested in natural history, especially conchology. In 1892 she entered upon her long service in the Buffalo Society of Natural Sciences, of which she became Director in 1899, finally retiring, after a connection of seventeen years, upon her marriage to Professor William Alanson Bryan in 1909. This long period was interrupted by several years given to study in the Academy of Natural Sciences of Philadelphia and the United States National Museum.

In 1899 the Conchological Society of Buffalo was organized by her, and a new period of local enthusiasm for the study of mollusks began. In 1906 Alfred University conferred the honorary degree of Doctor of Science. She was a member of the American Association for the Advancement of Science, the Conchological Society of Great Britain and Ireland, and various other scientific bodies.

Dr. Letson's publications relate chiefly to the mollusks of New York, the more extensive being a Check List of the Mollusca of New York, Bull. 341, N. Y. State Education Department, 1905; Post-Pliocene Fossils of the Niagara River Gravels, published in a Bulletin of the State Museum, 1901; a partial list of the shells found in Erie and Niagara counties and the Niagara frontier, Bull. Buffalo Soc. Nat. Sci., IX, 1909. At the time of her marriage to Professor Bryan, of the College of Hawaii, and her removal to Honolulu, she was working on a monograph of the New York Mollusca.

In Honolulu Mrs. Bryan engaged ardently in the collection of marine shells. Professor Bryan, who had before been chiefly known for his work on birds, added the mollusks to his other interests, and together, on many an island collecting trip, they amassed the largest collection of Hawaiian marine shells yet brought together.

For several years she had served as librarian of the College of Hawaii, a congenial task bringing many young people under her influence.

In 1917–18 Professor and Mrs. Bryan traveled in California and the East, spending several months at the Academy of Natural Sciences in studying Hawaiian shells. For the same purpose the museums of Cambridge and Washington were also visited.

Mrs. Bryan's gracious personality and sunny outlook, no less than the genuine love of nature which determined the course of her life, made her many warm friends who mourn her untimely death. H. A. P.

"Of Shoes — and Ships — and Sealing-wax"

An editor is always on the lookout for small articles that might interest his readers, and ones that will conveniently fill those white spaces at the end of an issue. The articles "of many things — whether of cabbages — and kings" are scattered throughout the early volumes of **The Nautilus**. Pilsbry started his career as a newspaper proofreader, and had an eye for newsworthy or amusing fillers.

CLAM-OROUS CROWS.—The following newspaper clipping, if true, shows that the amiable, inoffensive clams of the Northwest coast are having a hard time of it, and are entitled to the sympathy of all conchologists without distinction of age, sex or color:

"Scare-crows are now placed upon slate roofs in Victoria, B. C. The crows, which swarm on the beach and dig for clams, fly over the buildings and drop the clams on the roof, by this means breaking the shells and leaving the meat free to be eaten. In many cases, when the clams were dropped, the slate would be broken."

Such conduct on the part of the crows is certainly discreditable; they should be placed on the black-list.

> Ill fare the clams to hungry crows a prey,
> And brought to grief in such a crow-ill way.

The clam is probably *Saxidomus giganteus* Desh., quite common in the Vancouver region and the principal edible clam of both "Injuns" and white folks thereabout, and solid enough to break roof slates if not political ones, when dropped from a considerable elevation. *S. giganteus* is abundant between ordinary tide marks; it is great in soup; an excellent clam.—ROBT. E. C. STEARNS, *Los Angeles, Cal.*

A MAMMOTH LAND SNAIL.

In the *West American Scientist* for April, 1889, under the head of "A New Florida Bulimulus," follows the description of an alleged species of the group above named the dimensions of which are given as "length, 19 inches, diameter 8 inches." I don't believe that my esteemed friend Hemphill ever collected a land animal of the molluscan type quite as large as this. I wish that he had and I am sure if a beast of this size exists anywhere on the planet, it should when found be named for him, for I know of no man more worthy of such an honor. Let us return to the big Bulimus and consider its dimensions and what these figures mean:

Bulimus ovatus of Müller, a Brazilian species "attains the length of six inches and is sold in the markets of Rio." It has an egg an inch in length when hatched, say the size of a robin's egg. With this for a standard, the nineteen inch fellow from Florida may be expected under favorable circumstances and when not otherwise occupied to furnish eggs *three inches and upward* in length and of corresponding diameter. This looks like business, and here also is a hint in the way of a new industry. I was at one time slightly acquainted, with an old man, an alleged conchologist from the sunny land of France, of whom it was stated with much probability of truth, that he cooked common cowries in acid and bedeviled them in various ways, in the effort and hope to produce the beautiful Cypræa aurantia by an artificial process. His experiments were inspired not by scientific zeal but the lust of mammon. He did not succeed. His experiments rested on an imperfect ethical basis. But with the big bulimus as above, provided one could get enough to start the business and stock a small cochlearia or snail ranch, the business would be interesting scientifically and commercially and in no way *contra bona mores*. The proportions of the dividends compared to the profits of other kinds of business, might not be quite as large as the proportions of the big Bulimulus compared with the rest of his relatives.

But alas there are many incongruities and paradoxes in this world, and with this melancholy fact before us let us rest and find consolation, while dreaming of omelets and custards made of Bulimus eggs; and let us also in kindness overlook the infelicities of typographic errors and lapses of proof-readers.

R. E. C. S.

A TRAIN STOPPED BY SNAILS.—Mr. Laille, an engineer in the employ of the Tunisian Railway, writes in the *Dipeche Tunisienne:* "The train coming east from Suk-el-Arba last Thursday was two hours late for a very singular reason. The road was literally covered with snails, the wheels of the locomotive crushing these mollusks into a pulp, which destroyed all adherence and caused the locomotive wheels to skate, so to speak, in their places. We have seen flocks of locusts stop trains, but I think the fact that snails can stop a train is without a precedent. These snails are very general all through Tunis, especially during the rainy season; the smallest remainders of green on field or tree are covered with them, so much so that they appear like a bunch of grapes hung up, only that their white shells produce a curious effect."—*Phila. Record.*

THE WEIGHT AND SIZE OF SHELLS.

BY REV. HENRY W. WINKLEY.

With the assistance of Mr. D. E. Owen, teacher of Physics in Thornton Academy, the writer has weighed a few species of minute shells. The results are given as follows:

Twelve specimens of *Astyris lunata* from Wood Hole, Mass. weighed 0.095 gms. This would make one specimen weigh about 0.008 gm. Reducing this to avoirdupois weight we have one shell weighing 0.000282 oz.

The next example is *Cerithiopsis Greenii*—being the first of the species found in Canadian waters, i. e. from Prince Edwards Island. Ten specimens weighed 0.023 gm. or in ounces one specimen would weigh 0.000081 oz.

Two sets of *Odostomia seminuda* were compared. The one being, like the above, the first found at Prince Edwards Island. The others came from near Woods Hole, Mass. It was found that the Canadians weighed each 0.000048 oz. while those from Mass. weighed each 0.000105 oz. The difference in size is noticeable without weighing. This proves that Mass. is a better place to live than Prince Edwards Island. The most interesting of all is New England's conchological elephant, *Skenea planorbis*. The set weighed was found near Saco, Me. The average weight of a specimen is 0.000018 oz. At this rate it would require 56,700 to make an ounce, 907,200 to the pound, and a ton would require 18,144,000,000. At the rate of five cents each, a pound would be worth $45,360.00. I am sorry to say I cannot supply them by the ton, or pound.

After weighing, the writer became interested in size comparisons, and two species from the same region, i. e. Saco, were compared. The largest shell in my New England cabinet is *Mactra solidissima*, and the smallest *Skenea planorbis*. The Mactra weighs 17¼ oz. It would require 1,004,250 of Skenea to balance the one Mactra. The surface of the Mactra was reduced to a flat as near as possible, divided into small squares, and the Skenea was placed on the small square to estimate the comparative size. Dividing an inch into sixteen squares, Skenea would find room enough for 25 on each square, or 405 to the square inch. On the total surface of the Mactra (including both sides) there would be space enough for 30,000 individuals of Skenea to rest comfortably.

Estimate of the Number of Recent Mollusca

(70 years later, Dr. K. J. Boss at Harvard, after many months of investigations, independently came to the same conclusion. See Occasional Papers on Mollusks, *vol. 3, no. 40, pp. 81-135, 1971).*

The following interesting notes are taken from the Journal of Conchology, Vol. x, no. 2, pp. 35–42, April, 1901, "Conchology at the Dawn and Close of the Nineteenth Century" (The Presidential Address delivered by Mr. E. R. Sykes, at the Annual Meeting of the Conchological Society of Great Britain and Ireland, Oct. 27, 1900).

* * * "The close of the Nineteenth Century is, to use a commercial expression, a time to 'take stock,' and consider what progress has been made. It is with one of these forms of estimating our present position that I propose for a few minutes to concern myself, and especially with an endeavor to arrive at some idea of the actual number of species of recent mollusca which are now known to science. Any such estimate can but be approximate, but a survey of the most recent monographs enables one to form a fairly accurate conception.

"The classic starting-point for such a calculation, as indeed for all other systematic molluscan work, is the tenth edition of Linnaeus." His works contain roughly speaking about 700 species. This number gradually increased nearly every year, until "Dillwyn, in 1817, was enabled to enumerate 2,244; which we may divide into: *Cephalopoda*, 45; *Gastropoda*, 1,510; *Scaphopoda*, 15; *Pelecypoda*, 638; *Polyplacophora*, 36.

In the classic work of the brothers Adams (1853–58) we find the following: *Cephalopoda*, 197; *Gastropoda*, 12,604; *Scaphopoda*, 46; *Pelecypoda*, 4,258; *Polyplacophora*, 216.

Treating Paetel's well-known work (1888–1890) in the same way we get: *Cephalopoda*, 305; *Gastropoda*, 35,134; *Scaphopoda*, 137; *Pelecypoda*, 8,467; *Polyplacophora*, 439; or a total of 44,482 species.

"Hoyle's catalogue of the recent *Cephalopoda* in 1886, with addenda in 1896, contains 469. From the Zoölogical Record of 1897–9, we add eleven, and on an average, we may include four for 1900, making a total of 484 species.

In the *Gastropoda* the recent catalogue of the *Cyclophoridæ, Cyclostomatidæ,* and allies, by Kobelt and Moellendorff, "yields about 2,444 species, and if we add 48 species from the Zoöl. Record of 1899, and estimate a similar number for the 1900, we get 2,541." Since Paetel's list in 1888 (omitting the *Cyclophoridæ*, etc.), basing 1900 on a three years' average (682), there have been recorded 7,396, a total estimate of 43,021.

As to the *Scaphopods*, the most recent monograph, by Pilsbry and Sharp, yields 238; if we add the single one in the record of 1899, and another for 1900, we have 240 species.

For the *Pelecypoda* we have since Paetel's list (1890), basing as before, 1900 on a three years' average (142), 1,056; a total of 9,523 species.

"Finally we turn to the *Amphineura*. Here from Dr. Pilsbry's work we get: *Polyplacophora*, 540; *Aplacophora*, 33. Adding from the Zoöl. Record in a precisely similar manner we have to include *Polyplacophora*, 59; *Aplacophora*, 4; and we get a final total of 636.

"The next question which arises is, how far are the above totals trustworthy? On the one hand they are inflated by a mass of synonyms which still masquerade as species, while on the other hand they are reduced by a certain number of omissions. The only omission of any importance, however, will, I think, be found in the *Nudibranchiata*, of which the true total is, owing to the nature of the works consulted, unduly curtailed."

"Making a reduction therefore for synonyms and allowing for the above, I think a very fair approximation will be: *Cephalopoda*, 450; *Gastropoda*, 40,000; *Scaphopoda*, 220; *Pelecypoda*, 8,500; *Amphineura*, 600; or a grand total of 49,770—say 50,000 known species of recent mollusca." C. W. J.

ARGONAUTA FOUND ALIVE.—A living specimen of the Paper Nautilus, *Argonauta argo*, was found at Palm Beach, Dade Co., Fla., in April by Mrs. C. Rowland of Philadelphia. This handsome shell is over six inches in diameter. It is rare that living examples of this are found on our coast.

SEA GULL DROPS CLAM ON POLICEMAN.—The habit of sea gulls to carry clams in their talons to a considerable height and drop them on a hard surface to break, so that the bird can feed on the bivalve, nearly proved disastrous to Abe Loche, a former policeman of Atlantic City, N. J.

Loche was walking along the Boardwalk when one of the gulls flew high above him and dropped the clam directly on the man's head. He fell and had to be carried into a nearby drug store for treatment.—*Boston Globe.*

PLEUROCERA SUBULARE IN WATER-MAINS.

BY CHAS. T. SIMPSON.

The U. S. National Museum has recently received from the Hannibal Water Co., of Hannibal, Mo. (through Mr. Chas. T. Lewis), a number of dead shells of *Pleurocera subulare* Lea, taken from the mains and pipes of the company in that city.

Mr. Lewis states that they accumulate at the cocks and faucets, and seriously retard the flow of the water, putting the company to considerable expense to remove them; also, that none have been found in their reservoirs or settling-wells, and they have never seen them in the Mississippi River.

The specimens taken in the company's pipes are always dead, and are only found in a space of perhaps 12 to 15 blocks, and not all the pipes in this area are infested.

This species has been found as far west as the White River, Carroll Co., Arkansas, and in the Mississippi River at Davenport, from which localities specimens were obtained that are now in the National Museum Collection, though the range of this form is mostly to the eastward of these localities. It is probable that the eggs or very young entered the mains through the strainers and took up their abode in certain favorable localities in the pipes, where food was brought to them by the currents, or existed in abundance, and that a more careful search would disclose them in a living state in the service pipes.

To a Slug. (in alcohol.)

Hail, Limax!—clammy, slimy thing,
Poor houseless wretch, of thee I sing!
Though ended is thy earthly run,
Thy glory is but yet begun.
For Science, with obtrusive pride,
Will keep intact thy mortal hide
And suffer thee, for future gain,
In best of spirits to remain.

Oakland, Cal., Apr. 15, 1900.   H. H. BRUENN.

NEWSPAPER CONCHOLOGY.—The gloriously free daily press of this country does not often discuss scientific matters, but when it does, *facts* are apt to be mangled. The following clipping is not so bad: "It is generally supposed to be a sign of wet weather when snails go about without their shells. One species of snail never takes its walks abroad except when rain is at hand. Some climb trees two days before a down fall, setting upon the upper side of the leaves if a storm is to be of short duration, but taking shelter on the under side if it is to last some time. Still other snails turn yellow before rain, and blue when it is over."

NEWSPAPER CONCHOLOGY.—"One of the most beautiful shells found along our coast is that of a large snail which climbs certain trees and grows delicately fat on the young birds. The shell is as thin as tissue paper, oddly curved and almost as transparent as the finest glass. It belongs to the family of edible snails so prized as a delicacy on the coast of France, and if properly prepared makes a delicious dish. It is most abundant about New River inlet, where the slight shake of a tree about sunset will bring a shower of them to the ground. The breakage of a shell seems to be of little trouble to the snail—he repairs the damage and moves on."—*Jacksonville* (Fla.) *Citizen*.

A letter-writers' controversy that raged for several days in a New York newspaper brought out little known facts about the spelling of the "Muscle" part of Muscle Shoals. Philologists who contributed to Murray's exhaustive dictionary have mobilized no less than 26 English spellings for the mollusk called mussel. A writer of 1584 informed his readers that witches "can saile in an egge shell, a cockle or muscle shell." A traveler's book of 1681 reported that "The natives of Brasile use muscle shells for spoons and knives." In Glover's History of Derby (1829) mention is made of "a stratum of muscle shells." The poet Browning, in 1873, said that:

> "Granite and muscle shell are ground alike
> To glittering paste."

A Perfect Fountain Pen for 75 cents.

The only Fountain Holder practical with Steel Pens. We guarantee these pens to give perfect satisfaction. Sent postpaid on receipt of price. Address,
W. B. BENSINGER'S SONS, Tamaqua, Pa.

CORRESPONDENCE.

EDITORS NAUTILUS: In the remarks of our friend Hemphill, on pp. 139-140 of the April number, some of us are called to order for a phraseology which is not altogether agreeable to him and other workers on the Pacific coast, and may properly be modified in the interest of their feelings. I refer to expressions which refer to the habitual nomenclature of some shell under discussion, in the cabinets of collectors on the Pacific shores, as in error. I suppose I have been one of those whom he criticises; for being familiar with most of the West coast collections, lists and nomenclature, it has often seemed useful to refer to the name in common use on the other side of the continent, when in the course of monographic revision it has been found to be untenable.

But I should like to assure Mr. Hemphill and all others who have been displeased by such expressions that nothing was farther from my mind than to reflect on the care or desire for accuracy of West Coast workers. I have been of and among them so many years that I feel entitled to claim a place in their ranks, and a large part of the work I have done has been intended to assist them to the extent of my ability. No one is infallible, at least outside of the Vatican. No one can correct the errors of a nomenclature at one fell swoop, so to speak. I have named many thousands of shells for West Coast correspondents, and I have named many of them wrong. That is, I have given names which were at the time in current use, but which subsequent researches have shown to be untenable. If Methuselah was a conchologist, he probably did better toward the end of his career, and the heirs of his original correspondents profited thereby. But alas! these are degenerate days, and in forty years or so one does not ferret out all anterior mistakes. Therefore if one's gratuitous service does not prove infallible, and one's expressions not invariably happy, let our West coast friends hold fast to the theory of friendly intent, and believe we at the East mean to do our best by them every time. —WM. H. DALL.

FRIEND PILSBRY: Will you not suggest in the NAUTILUS that any conchologist travelling about the country should make it a point to call upon his brother collectors, if he goes near their homes? I hope no member if the A. A. C. who comes to Boston, will fail to visit me at Revere, only six miles away. Surely introductions are unnecessary in our little circle.—*Edward W. Roper.*

We heartily endorse this suggestion, friend Roper!

SNAIL PEST AT DENIYAYA, CEYLON. "The Kalutara Snails (*Achatina fulica*) at Deniyaya were found in such overwhelming numbers that at a meeting of the Morowak Korale Planters' Association held on March 7, 1925, the question of its eradication was discussed. This subject was referred to the Director of Agriculture who detailed me to make an attempt to eradicate the pest by the prompt application of systematic control measures.

The collection of snails was begun on May 19, 1925; snails and egg masses were hand-picked and destroyed by immersion in a 7 per cent solution of copper sulphate. Tamil cooly boys were employed for this work. The next stage was sanitation, and nearly 16 acres in Deniyaya village, where snails were found most, were clean weeded and all refuse burnt.

Over 60,000 snails and egg-masses were destroyed, and an expenditure of Rs. 203/42 were incurred in the operation (Year Book of the Department of Agriculture, Ceylon, 1926, p. 62).

PLOVER CAUGHT BY A PINNA—The article in the September number, "A Sora Caught by a Mussel," recalls a similar incident which I observed at New Pass, Sarasota Key, Fla. It was after a heavy storm, and there was a large number of *Pinna muricata* washed ashore. On my return to the boat after collecting some shells, I observed a Killdeer Plover that seemed very tame, but on a closer examination I found that it was caught by a *Pinna*. In this case it was caught by the bill. Its tongue was bleeding and the bill was indented by the sharp edges of the shell. It was with some difficulty that I removed the shell and let the Plover go rejoicing on its way.—E. J. Post.

SAD DEATH OF AN ORIENTAL BY HALIOTICIDE.—In the November *Popular Science Monthly*, Margaret Wentworth Leighton relates that while she was living in San Francisco, " A Chinamen went out on the rocks at low tide to gather some [Haliotis]. As he attempted to wrench one from its home his hand was caught between shell and rock, and so firmly held by the animal that he could not escape the rising tide and was drowned." West coast collectors should take warning. Don't fool with *Haliotis cracherodii* without having by you a crowbar or at least an ax, lest you should perish miserably like this child-like and bland Celestial.

Of Shoes – and Ships 273

Last week a coyote was found at Punta Banda, San Diego county, trapped by an abalone shell [*Haliotis cracherodii*]. The coyote had evidently been hunting for a fish breakfast, and finding the abalone only partially clinging to the rock had inserted his muzzle underneath to detach him, but the abalone closed down on him and kept him a prisoner.—*Weekly Bulletin*, San Francisco, May 17.

CURIOUS CHINESE USE OF SHELL-FISH.—The Chinese have been students of the habits of animals for many thousand of years, and the influences of this study have manifested themselves in their art and their architecture, so much so, that one can readily recognize the common form of their animal life through its resemblances to the objects and pictures with which we are familiar.

One of the most interesting is what is known as the "joss-shell." Every one has noticed the pearly luster of the bivalves of our rivers and ponds, fresh-water mussels, they are called. These mussels are lined throughout with the same kind of material as the pearl-oyster, and, indeed, pearls of value are often to be found in them. In China and Japan, these mussels grow to great size, in the latter country being oftentimes seven to ten inches in length, and in China, fully as large as a small saucer. The shrewd Chinese are aware that the pearly nacre is a protection of the animal, which has thus the smoothest of substances against its sensitive skin, and they know also that any grain of dirt or roughness will be quickly coated with pearl if it should lie under the mantle. They therefore catch the animal, and oblige it to make such designs as they desire. These are usually little josses, images of some one of the Chinese Gods, which are formed in clay and slipped between the mantle and the shell of the mollusk. The latter, as soon as it is put into the water again, begins to cover the model with a coat of pearl, and at some time, when the process has been carried far enough, the animal is killed and the shells preserved with their pearly josses and sold as curiosities. They are, however, very rare in this country, being on exhibition only in a few of the larger museums. It is said that upwards of one thousand of the Chinese made their living by this industry, and that they will, on order, insert in the shells models of the initals of any one's name, which, after a wait of a year and a half, will be ready for delivery.—*The Happy Thought*, July 15, 1895.

STRENGTH OF LIMPETS.

According to J. Lawrence Hamilton, M. R. C. S., the limpet is probably the strongest of known animals, excepting perhaps the *Venus verrucosa* of the Mediterranean Sea, which pulls 2,071 times its own weight when out of its shell. At Folkestone, Eng., Mr. Lawrence Hamilton found that the common sea shore limpet which weighs about half an ounce when deprived of its shell, required a force exceeding 62 lbs. to remove it from its powerful grip upon the rock, or 1,984 times its own dead weight. The superficial area of the base of the limpet experimented with measured 2.4 sq. inches. Mr. H. doubts whether the limpet's adhesive force has anything to do with the question of atmospheric pressure. A curious illustration of the limpet's strength is given by another naturalist. On a warm dry day in summer, on the Northern Coast of Scotland, a hare approached a limpet and endeavored to moisten its tongue by contact with the watery looking flesh of the latter; instantly, the limpet closed on to a rock pinning the hare fast by the tongue and holding it until the animal was caught by the observer of the occurrence.

The life story of Sir Marcus Samuel, who has purchased from the Earl of Berkeley for the sum of $25,000,000 a parcel of the fashionable residential section of London, known as Berkeley Square, furnishes one of the real romances of the business world.

Sir Marcus, in his early days, kept a little shop in one of the poorer quarters of the British metropolis, where he made and sold for a shilling or two ornamental boxes made of shells from the seashore. Later he invested his savings in oil, made money and started a company called the "Shell," thus identifying his big new venture with his original struggling business. Since those days he has accumulated a fortune of many millions and has been honored with a baronetcy. And all from selling shells from the seashore—mixed with an abundance of brains and energy.—*Washington Evening Post.*

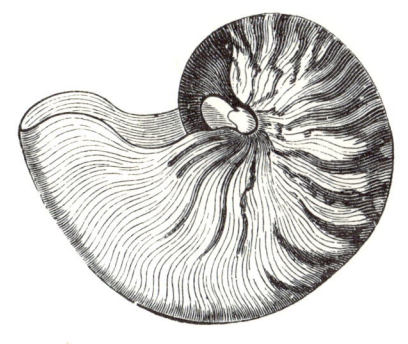

Salute to The Nautilus

by Henry Dodge

This poem was written in 1959 by Henry Dodge, a New York attorney and amateur conchologist. Dodge lived in Scarsdale, New York, and during the latter part of his life wintered in Nassau, Bahamas. He took a keen interest in early conchological literature, and was a prodigious writer on the identity of Linnaeus' species. The poem was first published in 1960 in volume 1, number 4 of the Proceedings of the Philadelphia Shell Club.

For eight and sixty fruitful years
The Nautilus has flourished.
For years our conchological
Appetite it's nourished.
Its pages are a Hall of Fame —
Mollusca's men of note.
So here's an alphabetical
Review of those who wrote:

For A there's old-timer "Helicidae" ANCEY.
(He'd write about any old genus you'd fancy.)
And R. Tucker ABBOTT has shown what ails
Those fellows who play with the Schistisome snails.
To a brand-new position he's just fallen heir,
For Tucker now graces the Pilsbry Chair.
B is for BARTSCH who wrote monographs daily,
And BINNEY and BEQUAERT, two BERRYS and Baily;
A big crop of BAKERS — Fred, Frank C., and Horace,
For Land and Fresh water a real B-sharp chorus.
And don't forget BORMANN, the whole tribe of BURCHES,
Doc BALES' and Ted BAYER'S vast submarine searches.
C stands for CLENCH whom the M.C.Z. took;
His "Johnsonia's" really a whale of a book;
And "Partula" CRAMPTON and COCKERELL and CLAPP,
And Montagu COOKE, who has left such a gap.
The D is for DALL, our Master for years.
He stood all alone in the ranks of the Peers.
I needn't cite other men under the D's
Lest we might miss the forest because of the trees.
The E recalls EYERDAM's boreal beat
And EDSON and EMERY down in St. Pete.
And F stands for FERRISS. What tales he could tell us!
And "Pliocene" FARGO who works in Pinellas.
The first roamed the country where dwelt the Apache,
While Fargo hunts fossils in Caloosahatchee.

Salute to The Nautilus

For *G* we have GRATACAP, GRANT (Hoyt Gale's pardner);
And GOODRICH and GREGG, and for gals, Julia GARDNER.
In *H* there's HAAS, HERTLEIN and HEDLEY to pick;
The far-off Antipodes *his* bailiwick;
Two HENDERSON chaps, whose output was terrific,
And old Henry HEMPHILL out on the Pacific.
The *I* stands for IREDALE, Seer of Australia;
(If you should contest this he'd probably jail ya)
His Cypraeid genera leave me quite numb.
He out-Schilders Schilder, and that's going some!
In *J* there's a standout — C.W.J.
We won't see his equal for many a day;
A scholar, a teacher, yet never pedantic,
And, Boy! What he did for the Western Atlantic!
For *K* it was KEEP in the West Coast procession;
KENNARD the taxonomist — thankless profession!
L stands for LERMOND, old Maine's beloved sheller,
And LOWE who's the West Coast's most famed storyteller.
And don't forget LYMAN; "It's a great day!" he's yelling,
For Frank and his family just live to go shelling.
In *M* there are MORRISSON, MARSHALL and MORSE,
McLEAN and McGINTY, and MELVILL of course;
And South Carolina's old William MAZYK,
And Doc MOORE, who gives shells to the nervous and sick.
For *N* there's another Down-easter in Maine,
For NYLANDER cropped up again and again.
And *O* has the OLDROYD's, whose works we acclaim,
And OLSSON, and ORTMANN of Naiadae fame.
P stands for PILSBRY, old true and tried,
The "Nautilus" editor, mentor and pride.
All other P-names I am willing to pass,
For no other sheller was quite in his class.
In *Q* "Tapes" QUAYLE was another first-rater.

(We might have another — if QUOY had lived later.)
And R gives us REHDER, a top-notch Curator,
Who knows all the mollusks from Pole to Equator;
From the list of new species now laid at his door
You'd think we were hearing Gmelin once more!
There's RICHARDS and ROBERTS, and ROPER from Maine.
The old Pine Tree State's in the roster again.
In S there's a plethora — with SMITH we begin —
Allyn, Maxwell and Herbert, E. A. are all in;
(And a virtual homonym, not quite a fit,
For in Vol. number 60 there's even a SMIT.)
We've STRONG and we've SCHWENGEL who helped H.A.P.
And SHIMEK, and SIMPSON, a prince, you'll agree,
And SPICER and SCHILDER, the king of the splitters,
A genera builder who gives me the jitters.
And R.E.C. STEARNS so prolific and wise,
And STERKI, who had those real microscope eyes.
T stands for THAANUM, the Islands' real Dean
"Cone" TOMLIN, whose list is the finest we've seen,
And C. de la TORRE, a most honored guest
And well-beloved neighbor, who's gone to his rest.
Just one for U, Oh, Gentle Reader,
It's UTTERBACH the mussel-breeder.
But when we come to the letter V
We've several names for you to see:
First, there's VANATTA who had lots of "savvy."
His range was from Zion clear down to Gonave.
And H. VAN DER SCHALIE, a toast of Ann Arbor's,
VAN HYNING, who's screened most of Florida's harbors
In W, WHEELER and WINKLEY and WILLET,
Though the list is so long, there are good men to fill it.
There's WEBB the anatomist, Berlin H. WRIGHT;
On the *Unio* muddle he provided some light;
And WOODWARD (B.B.) who from Old England hails,

And wise Bryant WALKER, Law's gift to the snails.
When we come down to X there is no one to cite,
But I'm sure that the future will give us some light.
Let's say "the unknown" who will take up the quest
When all the old Masters have passed to their rest.
The Y's are not numerous, though YOKUM's on file,
And YATES from the West Coast wrote once in a while,
And though TING CHIEN YEN's not a writer prolific
We welcomed this lad from across the Pacific.
In Z there's a singleton, ZETEK by name,
He wrote on *Drymaeus*; from Balboa he came.

At last I have finished; the alphabet's done.
I've had to omit and I've praised and poked fun.
Let's hope, as the years pile up score upon score,
There'll be the old "Nautilus" just as before;
And I know our successors and those who inherit us
Will revere H.A.P. as Docanus Emeritus.

Finest Specimens and Rarest Species
Of N. A. Helices, UNIONIDÆ and STREPOMATIDÆ,

Clean and perfect, For Sale at reasonable prices. Send list of desiderata which will be priced. Wish to buy for cash *any species* not in my Collections.

R. M. WETHERBY
Magnetic City, Mitchell Co., N, C.

Marine shells, Inveterbrate Cretaceous Fossils, Minerals, Polished Agates, Corals, and a few species of land and fresh water shells for exchange, want Marine shells and publications on Conchology. **HOMER SQUYER, Mingusville, Mont.**

THE NAUTILUS.

ADVERTISING RATES.

Advertisements will be inserted at the rate of $1.00 per inch for each insertion in advance. Smaller space in proportion. A discount of 25 per cent. will be made on insertions of six months or longer.

SPECIMEN TRAYS
FOR CABINETS.

Jesse Jones & Co., PAPER BOX MAKERS, 615 COMMERCE STREET, PHILADELPHIA.

RECENT SHELLS.

HELIX FULTONI, *Godwin-Austen.* $1.20.

Hugh Fulton has just acquired the grand collection of land and fresh water shells formed by the late MONSIEUR ARTHUR MORELET, probably the largest collection ever formed by an amateur, and containing nearly 10,000 species.

HUGH FULTON (Conchologist),
216 KING'S ROAD, LONDON, S. W.

JAMES W. QUEEN & CO.,
1010 CHESTNUT ST., PHILADELPHIA.

Microscopes and Supplies, Magnifying Glasses, Insect Pins, Sheet Cork, Plant Presses, Collecting Cases, Etc.

Send for Complete CATALOGUE **B.**, also Sample Copy of the MICROSCOPICAL BULLETIN.

LIBRARY OF DAVIDSON COLLEGE

Books on regular loan may be checked out for **two weeks**. Books must be presented at the Circulation Desk in order to be renewed.

A fine is charged after date due.

Special books are subject to special regulations at the discretion of the library staff.

MAY 24.1984